Poles and Zeros

in electrical and control engin

Poles and Zeros
in electrical and control engineering

R. J. Maddock
Principal Lecturer
Department of Systems and
Communication Engineering
Southampton Institute of Higher Education

CASSELL

Cassell Educational Limited,
Artillery House,
Artillery Row,
LONDON, SW1P 1RT

First published 1982
Reprinted 1987

ISBN 0 304 314609

Printed and bound in Great Britain at The Bath Press, Avon

Preface

Certain branches of engineering often use complex frequency or *s* domain analysis. This is particularly common in the study of electrical filter networks, operational amplifier circuits and control systems but it will also be encountered in other branches of electrical and electronic engineering. Books and papers on these subjects are freely illustrated with pole zero diagrams and root locus plots. These are most useful in appreciating system behaviour and also in the synthesis or modification of systems. They are both concerned with complex frequency analysis and unless the reader has been introduced to this subject, the article in question may be quite incomprehensible.

Courses for technician engineers and higher technicians in electrical subjects and control engineering usually include material that uses Laplace methods to solve transient problems introductions to the concepts of transfer functions and pole and zeros. This work may then be extended in various subsidiary courses. There are many books aimed at final degree courses and post-graduate study which include the required range of material; these tend to be too academic and their treatment is not suited to their readers.

This book is intended for the following readers: students following TEC Higher Certificate or Diploma programmes in electrical, electronic or control engineering; students attending courses leading to a Higher National Diploma in electrical and electronic engineering; undergraduates reading for degrees in electrical engineering, electronic engineering or control engineering; and practising engineers and higher technicians who encounter complex frequency and poles and zeros in their reading.

The principle aim of this book is to treat the use of complex frequency in the same way as many books on a.c. circuit theory use phasors of j notation. The development of a.c. theory from the solution of differential equations is frequently either ignored or dispensed with as quickly as possible. The techniques of phasor or j notation are then considered in depth and applied to the solution of many circuits and problems. In the same way, the lack of a deep understanding of Laplace transformation does not prevent analysis in the *s* plane or an appreciation of the relationships that can be found. Higher technicians can appreciate these relationships provided they are developed in a logical sequence and they are not buried in mathematical notation which the reader has no feeling for. Many engineers may also find this practical approach helpful in the understanding of electrical circuits and control systems.

In this book, the following steps are used to arrive at this objective.

1. The development of the use of Laplace transform methods as a 'tool' for the analysis of transients in electrical circuits. The mathematical expertise required is, for the most part, limited to algebraic manipulation.

2. Familiarization with the concept of transfer function and the representation of transfer functions by their poles and zeros.
3. Demonstration of simple techniques by which the pole zero diagram may be used to predict or estimate the transient response of a circuit or system.
4. Demonstration of simple techniques by which the pole zero diagram may be used to predict or estimate the steady-state frequency response of a circuit or system.
5. Discussion aimed at giving the reader a 'feel' for the relationship between the pole zero diagram and the resulting circuit behaviour.
6. Introduction to root locus plotting to aid investigation the variation of the behaviour resulting from the change in a particular component or parameter.
7. Demonstration of root locus techniques for a range of practical problems.

Throughout the book, each new concept or technique is illustrated by fully worked numerical examples. At the end of each chapter, a large number of graded examples for further practice are included. Answers are provided with each example, in a form suited to the type of question. For example, where the required answers can be obtained 'by inspection' from a large number of question parts, these answers are simply listed without numeration. On the other hand, problems requiring calculation and analysis have the answers fully tabulated, and those requiring the construction of a graph have the essential points on the graph listed so that the solution may be easily checked. Certain techniques of network analysis which may be unfamilar to some readers have been discussed in appendices to avoid interruption of the main argument.

Much of the work outlined above uses graphical techniques requiring the determination of the lengths and angles of phasors between two points on a graph. These may be *measured* using ruler and protractor, or alternatively, the combined measurement can be found by using a spirule, an instrument which has been designed specifically for this type of work. Another method is to calculate the required lengths and angles using any scientific calculator. This process will be quicker when dealing with points along the real or imaginary axes, but for complex points, measurements will probably be quicker. A programmable calculator (with sufficient program steps and memories) can be programmed to perform all the tests and measurements required for practical systems. The author has found this technique to be the most convenient and this has been applied throughout the book.

Robin J. Maddock, Southampton

Contents

1

Time and Frequency Domains

1.1 INTRODUCTION

In order that the behaviour of physical systems may be understood, the action of the components and variables of the system must be described in a precise manner. Such descriptions are usually mathematical equations or relationships. These equations are then further combined to include any interaction between the system components. In many cases, the resulting sets of equations are extremely complex and even if a single numerical answer can be obtained, a general understanding of the system behaviour is not at all obvious. In some cases, a more convenient description of the system can be obtained by the process of *transformation*. The principal objective of this book is to show how the transformed world of complex frequency and the s plane can be used to understand and predict the behaviour of certain classes of physical system. A useful understanding can be obtained without a deep knowledge of the mathematical principles behind the transformation. To go even further, the relatively simple mathematical manipulations can be employed to predict system behaviour without any knowledge of the background. However, most people find it advantageous to have some general ideas on which to base the techniques they are employing.

In the first part of this chapter, a general background to the concept of transformation is discussed. As with the remainder of the book, the main vehicle for discussion is that of electrical systems, variables and components. These have been chosen as a wide variety of concepts and problems can be illustrated with such systems. The general ideas and techniques demonstrated can, however, be applied to other physical systems and are particularly useful in the field of control engineering. For this reason, wherever possible throughout the book, examples involving control systems have been included.

In the following sections of this chapter, the nature of complex frequency is discussed and the techniques for analysis are illustrated with problems involving electrical and other physical systems.

1.2 THE RESPONSE OF ELECTRICAL NETWORKS TO SIGNALS

Electrical networks consist of the passive elements, resistors, inductors and capacitors, interconnected in various arrangements, with or without active devices which can amplify the signals applied to them. The electrical variables that occur in the networks are voltages or potentials, v, and currents, i, where v and i are the instantaneous values of voltage and current respectively. The components (R, L and C) are defined in terms of the two variables, as listed in Fig. 1.1 below. For example, if a current i is flowing in an inductor, the voltage across the inductor is given by $v = L\dfrac{di}{dt}$. Alternatively, we can say that any physical object for which the voltage is proportional to the rate of change of current can be called an inductor; the constant of proportionality is then the inductance L of the object. Similarly, if the current is proportional to the rate of change of voltage, we have a capacitor C. If the voltage is simply proportional to the current itself, then the component is a resistor.

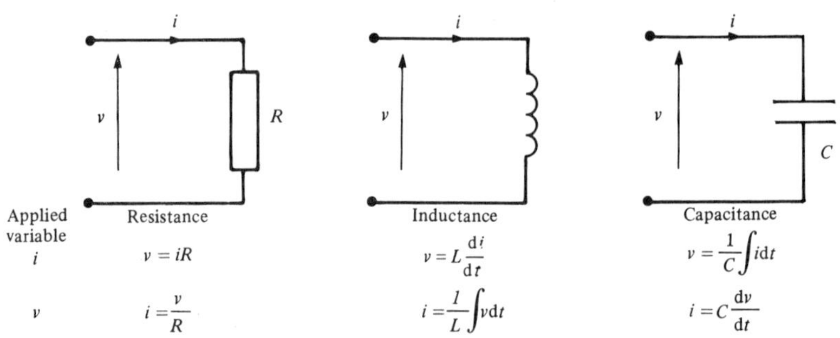

Fig. 1.1 The fundamental relationships for passive electrical components.

When the values of the variables v and i are related in terms of the components, time becomes an important factor. Currents and voltages must therefore be expressed as functions of time $i(t)$ and $v(t)$ respectively. Applied signals will also be functions of time; usually voltages, but occasionally current signals must be considered. The response of the network to these signals will be a voltage or a current in some other part of the network. This too will be a function of time, $r(t)$. Fig. 1.2 shows some typical signals as the waveforms that are obtained when the variable is plotted as a graph for increasing time t.

The response waveform may also include ramps, sinusoids and d.c. levels, but other functions of time such as exponentially decaying voltages will also appear.

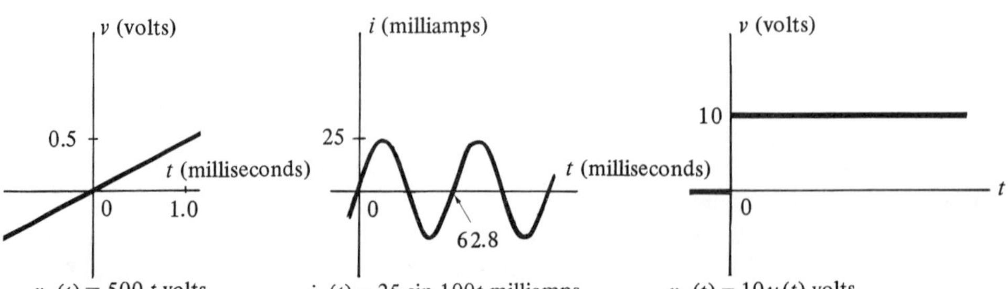

Fig. 1.2 Waveforms for time domain signals.

1.3 TIME DOMAIN ANALYSIS

As described above, the nature of the system components requires us to express all our variables as functions of time. This is also true of other physical systems such as bodies having mass being moved against frictional forces with the aid of elastic springs. In such a system, either the position or the velocity will be expressed as a function of time. Analysis of all such systems using time dependent relationships is referred to as time domain analysis.

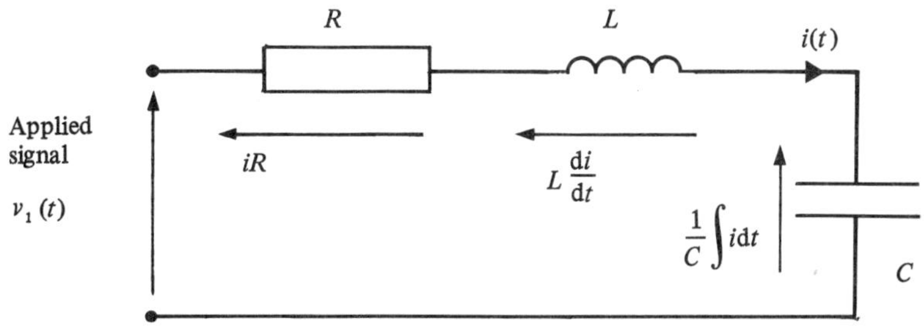

Fig. 1.3 Time domain analysis of a series *RLC* circuit.

Returning to our electrical networks, we can write equations using Kirchhoff's laws to obtain relationships for the complete network. Fig. 1.3 shows a simple network with an applied signal: the required response is the current flowing in the capacitor, i.e. the circuit current.

The system equation is:

$$v_1(t) = iR + L\frac{di}{dt} + \frac{1}{C}\int i\,dt. \qquad \ldots(1.1)$$

Differentiating this expression results in a second-order differential equation which could be solved using traditional methods. This circuit is, however, very simple, yet such a solution would be considered difficult or long-winded by many students. We frequently encounter networks with many more components, interconnected to form two or more loops. The resulting system equations will be a set of simultaneous differential equations which are very time-consuming (if not impossible) to solve.

1.4 FREQUENCY DOMAIN ANALYSIS

The problem of solving differential equations can be overcome by limiting the problem to a reduced class of functions of time for which simple differential relationships exist. The first class of this type is d.c. for which all voltages and currents do not vary with time and are constant. The differential terms disappear resulting in no drops in voltage across inductors and no current in capacitors (see Fig. 1.1). A more important class is steady-state a.c. This form of variable is applicable to most power circuits and to many communication systems (at least to a first approximation). If the signal is of the form: $v(t) = \hat{V} \sin \omega t$, then all the currents and voltages in the system will be sinusoidal and at the same angular frequency ω. Consider, for example, an inductor carrying a current: $i(t) = \hat{I} \sin \omega t$; the voltage across the inductor is given by:

$$v(t) = L\,\frac{\mathrm{d}i(t)}{\mathrm{d}t} = L\hat{I}\frac{\mathrm{d}}{\mathrm{d}t}\sin\omega t\,. \qquad \ldots(1.2)$$

Differentiating sin ωt.

$$v(t) = L\hat{I}\omega\cos\omega t = \omega L\hat{I}\sin(\omega t + 90^\circ)\,.$$

The same term, sin ωt, which defined the current appears in this result, together with a phase advance of 90° and a constant of proportionality ωL which is a *function of frequency* ω.

We can now transform into the freauency (jω) domain and write:

$$V = \mathrm{j}\omega LI, \qquad \ldots(1.4)$$

where the implication is that V, I represent rms values and that the j is indicative of *sinusoidal* operation.

This is the familiar a.c. theory form and in a similar way for capacitors and for resistors, we write respectively:

$$V = \frac{I}{\mathrm{j}\omega C}, \qquad \ldots(1.5)$$

and

$$V = IR. \qquad \ldots(1.6)$$

By the use of these jω forms, the writing of Kirchhoff equations no longer involves differentials and the results can be manipulated algebraically (remembering that $\mathrm{j}^2 = -1$). To obtain a numerical solution, we have to substitute a numerical value for ω. The resulting response will be of the form:

$$V_o = V\angle\phi. \qquad \ldots(1.7)$$

If required, this result can be transformed back to the time domain to give:

$$v_o = \hat{V}\sin(\omega t + \phi). \qquad \ldots(1.8)$$

Finally, a value of instantaneous voltage can be obtained for a specified time by substituting the required t value. However, the instantaneous voltage is rarely required; in practice, the frequency domain form $V\angle\phi$ or $I\angle\theta$ is usually satisfactory. For example, when a domestic electrical supply is connected to an electric fire, the applied voltage is 340 sin 314t V and the resulting current would be perhaps 17.7 sin (314t – 2°) A, (the 2° assumes that the element is slightly inductive). Most people would be quite at home with the frequency domain form of 240 V, 12.5 A lagging by 2°.

The whole process of transformation and frequency domain analysis can be illustrated diagrammatically, as in Fig. 1.4.

The signals to be applied to the network are not necessarily simple sinusoids. By the principle of superposition, if a signal voltage can be shown to be the sum of a number of other voltages, the response will be the sum of the responses due to each component voltage taken separately. The Fourier Series theorem shows that any continuous, periodic wave can be represented by the sum of a number of sine waves at frequencies which are multiples of that of the original periodic wave. The sum is, in fact, that for an infinite but convergent

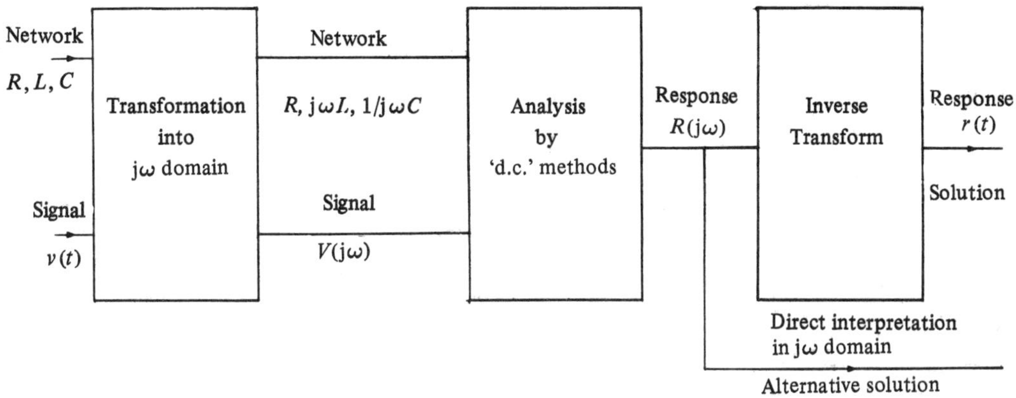

Fig. 1.4 The process of frequency (jω) domain analysis.

series. If the signal is a square wave as shown in Fig. 1.5a, the sum of the first seven or eight terms will give a reasonably good approximation to the required signal as shown in Fig. 1.5b. The rectangular wave shown in Fig 1.5c is also continuous and periodic, and has the same frequency as the square wave. It is not surprising, however, that to find a reasonable approximation in this case, many more terms would be required and the individual component voltages would be much smaller. If this concept is extended to the limit, the frequency becomes zero and there is a single pulse of voltage. Now an infinite number of infinitely small sinusoids must be summed to obtain the result. This sum is given by the Fourier transform integral:

$$V(j\omega) = \int_{-\infty}^{+\infty} e^{-j\omega t} v(t)\,dt \,. \qquad \ldots (1.9)$$

These Fourier methods can be used to find the response of a system to periodic waves, but they cannot be used to transform signals which are zero for time less than 0 and have finite values for time tending to infinity. This includes any signal that is 'switched on' at the beginning of the analysis. To investigate the transient response after switching, we shall have to use another form of transformation.

The Fourier integral (1.9) is an infinite sum of constant amplitude or undamped sinusoids. An alternative, the Laplace integral or transform is an infinite sum of exponentially decaying or damped sinusoids. The Laplace transform is given by:

$$F(s) = \int_{0}^{+\infty} e^{-st} f(t)\,dt, \qquad \ldots (1.10)$$

where s is a complex frequency $\sigma + j\omega$.*

1.5 COMPLEX FREQUENCY *s* DOMAIN ANALYSIS

Transformation into the complex frequency or s domain will allow us to analyse the

* If $\sigma = 0$, then the Laplace transform of (1.10) reverts to the Fourier transform of (1.9). Thus the Laplace transform is a generalization of the Fourier transform.

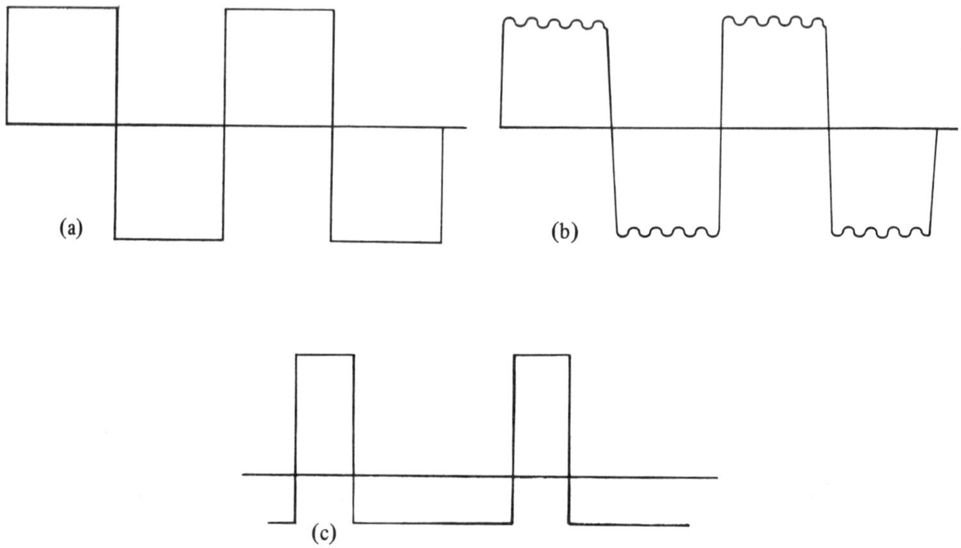

Fig. 1.5 Examples of continuous periodic waves.

response of systems to switched input signals and the result will include both the transient and the steady-state response. Thus if only the steady-state response is required, the transient component can be ignored by allowing the time t to become large. Thus the results obtained in the s domain *include* the familiar d.c. and a.c. theory. The mathematics involved does not require us to evaluate integrals of the form given in expression (1.10) as tables of all the transforms required are readily available. In addition, should the solution of differential equations be required (particularly in non-electrical systems), the Laplace method is just as powerful.

Before considering the techniques of analysis in the s domain, it will be useful to examine the nature of s in more detail. Dimensionally, s is a frequency having units proportional to the reciprocal of time. It is a complex quantity representing both an oscillating or sinusoidal quantity ω and an exponentially decaying (or growing) quantity σ. Points representing these two components can be plotted on the complex s plane shown in Fig. 1.6. The axes for this diagram are $\pm\sigma$ and $\pm j\omega$ and waveforms for signals represented by points (or pairs of points) are shown in the insets. In each case the signal is given by e^{st} which is equal to unity when $t = 0$. This form can be expanded into:

$$e^{st} = e^{(\sigma + j\omega)t} = e^{\sigma t} . e^{j\omega t}. \qquad \ldots(1.11)$$

Consider first a value of s where $\omega = 0$ and $\sigma = -\sigma_2$. The term $e^{-\sigma_2 t}$ is exponentially decaying from unity at $t = 0$ with a time constant $1/\sigma_2$. At $\sigma = -\sigma_3$ and $-\sigma_1$, the time constants for the decay are shorter and longer respectively as shown in the diagram. At $s = \sigma = 0$, there is no exponential decay; thus this point represents a d.c. or constant signal for $t > 0$. A positive value of $\sigma = \sigma_4$ represents an exponential growth which will tend to infinity. This is an unstable condition and in practice the system limits the growth by some form of saturation or self-destruction.

Now consider values of s which are imaginary with no real parts. Here, conjugate pairs of points must be taken since the exponential form for a cosine is given by:

$$\cos \omega t = \frac{e^{j\omega t} + e^{-j\omega t}}{2} \,. \qquad \ldots (1.12)$$

Referring again to Fig. 1.6, we can see that the pairs of points at $\pm j\omega_1$ and $\pm j\omega_2$ represent steady or undamped cosine waves having angular frequencies of ω_1 and ω_2. Once again, as ω tends to zero at $s = 0$, we return to the d.c. condition.

A complex conjugate pair of points at $-\sigma_1 \pm j\omega_2$ shows a signal oscillating at ω_2 rads^{-1} and decaying with a time constant $1/\sigma_1$. Other such pairs in the left half of the s plane will show oscillating frequencies according to the value of $\pm j\omega$, decaying at a rate dependent on $-\sigma$. If $|\sigma|>|\omega|$, the rate of decay will be such that the oscillatory nature is barely apparent; this is shown with the points $-\sigma_3 \pm j\omega_1$.

Finally, a complex conjugate pair of points in the right-half plane, $+\sigma_4 \pm j\omega_2$ represents an oscillatory signal with exponential growth to infinity. In general, if signals represented by points in the right-half plane occur in a system, the system is said to be unstable.

1.6 THE TECHNIQUES OF *s* DOMAIN ANALYSIS

The steps for s domain analysis of electrical networks are similar to those illustrated for the $j\omega$ domain in Fig. 1.4. There may, however, be information additional to that provided by the network, or system, and the applied signal. Analysis will start at $t = 0$, at which time, the variables may either have zero value or a finite value resulting from conditions prior to $t = 0$. Any such finite values for the variables are known as the initial conditions for the system. Analysis of non-electrical systems may also involve initial conditions and examples of such systems are considered at the end of the chapter.

The general procedure for the s domain analysis of electrical networks is illustrated in Fig. 1.7.

The first step, transformation into the s domain, is accomplished by the use of Laplace transform tables. An abbreviated set of Laplace transforms is given in Table 1.1. This contains all the transforms that are required for the work in this book. Signals can be transformed directly, but components and initial conditions are treated first so that the electrical network can be redrawn in the s domain (in the same way as a.c. networks are drawn showing resistance, reactance and impedance). The resulting s domain network is then analysed using all the familiar d.c. methods; Kirchhoff's laws, mesh and nodal analysis, Thevenin's theorem, etc. The response $R(s)$ is a voltage $V(s)$ or a current $I(s)$, all of which are s domain transformed variables. Initially, this response is transformed back into the time domain by means of algebraic manipulation and comparison with the table of transforms. The resulting time domain response $r(t)$, includes two components, $r_{tr}(t)$ the transient component and $r_{ss}(t)$ the steady-state component. These results are often best understood by graphical representations which can be achieved by sketching methods to be illustrated by examples. In later chapters, we see how the response $R(s)$ can be understood in either the time domain or the $j\omega$ domain without having to complete the inverse transform.

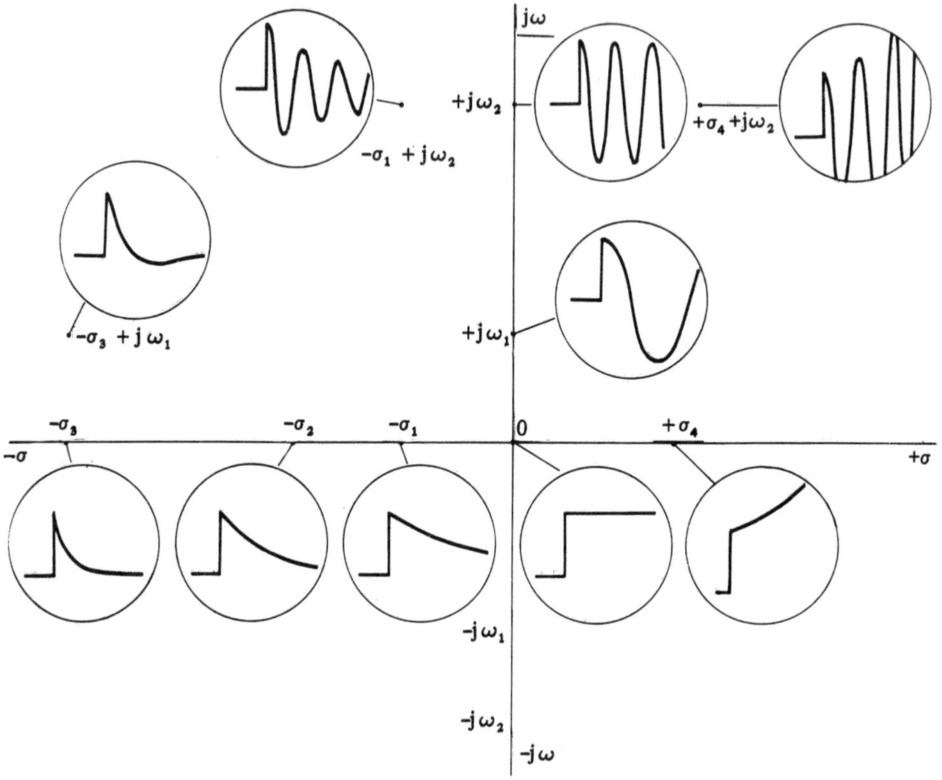

Fig. 1.6 Waveforms for complex frequency signals.

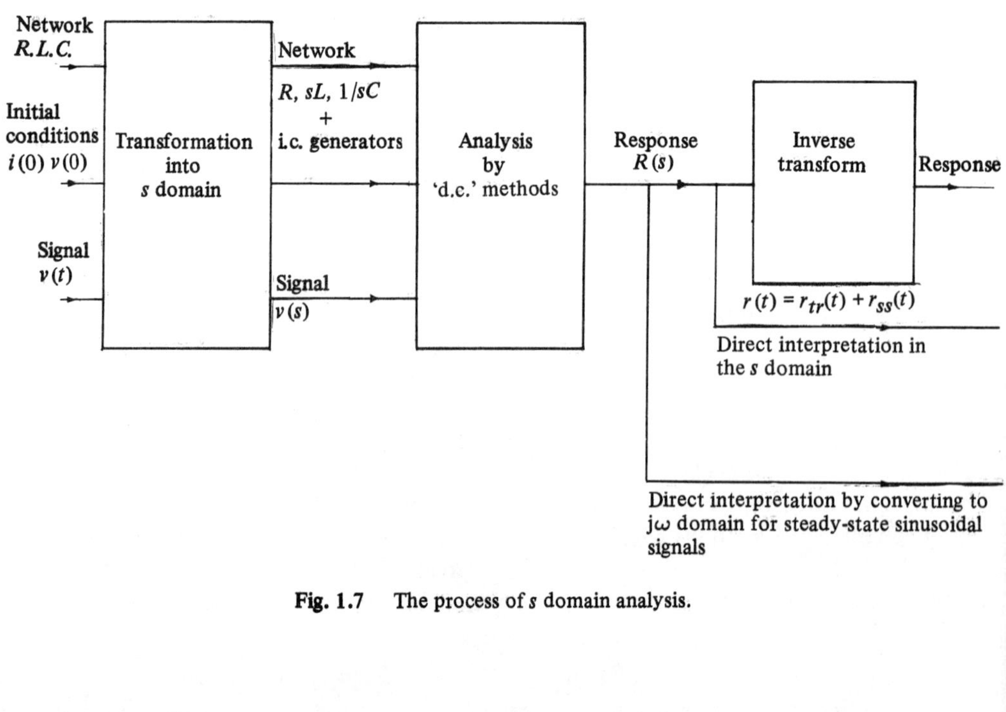

Fig. 1.7 The process of s domain analysis.

Table 1.1 is divided into three sections, the first of which is a list of the transforms of variables that will be encountered as signals or response terms in this book. The second section is concerned with the differentials and integrals of functions in the s domain, where the transform $F(s)$ of the particular function is already known. These are used to transform the electrical and initial conditions into their s domain equivalents. They are also used in the solution of differential equations where these have been written from first principles for a particular system. The two theorems in the final section are used in the estimation of responses in Chapter 3.

TABLE 1.1 LAPLACE TRANSFORMS

	Signals or responses	f(*t*)	F(*s*)
1	*Step or d.c.*	$au(t)$	$\frac{a}{s}$
2	*Ramp*	at	$\frac{a}{s^2}$
3	*Parabola*	at^2	$\frac{2a}{s^3}$
4	*Exponential decay*	ae^{-bt}	$\frac{a}{s+b}$
5	*Critical damping*	ate^{-bt}	$\frac{a}{(s+b)^2}$
6	*Sine wave*	$a \sin \omega t$	$\frac{a\omega}{s^2+\omega^2}$
7	*Cosine wave*	$a \cos \omega t$	$\frac{as}{s^2+\omega^2}$
8	*Damped sine wave*	$ae^{-bt} \sin \omega t$	$\frac{a\omega}{(s+b)^2+\omega^2}$
9	*Damped cosine wave*	$ae^{-bt} \cos \omega t$	$\frac{a(s+b)}{(s+b)^2+\omega^2}$
10	*Impulse*	$\int_0^\infty \delta(t)\, dt = 1$	1
11	*Differentials*	$\frac{df(t)}{dt}$	$sF(s) - f(0)$
12		$\frac{d^2 f(t)}{dt^2}$	$s^2 F(s) - sf(0) - f'(0)$
13	*Integration*	$\int_0^t f(t)\, dt$	$\frac{F(s)}{s}$
14	*Initial value theorem*	$\lim_{t \to 0} f(t) = \lim_{s \to \infty} sF(s)$	
15	*Final value theorem*	$\lim_{t \to \infty} f(t) = \lim_{s \to 0} sF(s)$	

1.7 TRANSFORMATION OF COMPONENTS INTO THE *s* DOMAIN

The three electrical components that are transformed are inductance L, capacitance C and resistance R. Fig. 1.8 shows an inductor in the time domain including the initial condition of a current $i(0)$ flowing at $t = 0$. From fundamental principles,

$$v(t) = L\frac{di(t)}{dt}\ . \qquad \ldots(1.13)$$

If the transformed variables are $V(s)$ and $I(s)$, transform rule 11 in Table 1.1, for a differential, can be applied.

$$V(s) = L(sI(s) - i(0))\ , \qquad \ldots(1.14)$$

$$= sLI(s) - Li(0)\ . \qquad \ldots(1.15)$$

Expression (1.15) is the sum of two voltages and can be represented by the drop in voltage across two components in series as shown in Fig. 1.8b. The first voltage is proportional to the current flowing $I(s)$ and is therefore the voltage across an impedance of value sL. This can be drawn as a block, but it may be convenient to draw it as an inductor as a reminder that it is an inductive impedance (not reactance; the concept of reactance does not occur in the s domain since magnitude *and* phase are implicit in s). The second voltage, $-i(0)L$, cannot be represented by an impedance as the current $i(0)$ is not flowing in the branch. The alternative is a voltage generator, in the direction shown, of $-i(0)L$ volts. It is more convenient to show this initial condition generator with the opposite sense or direction and a value $+i(0)L$. We can then note that *the direction of the initial condition generator is the same as that of the initial current in the time domain circuit.* The final result is shown in Fig. 1.8c. Note that to represent an inductor in the s domain, both the impedance sL and the initial condition generator *must* be included.

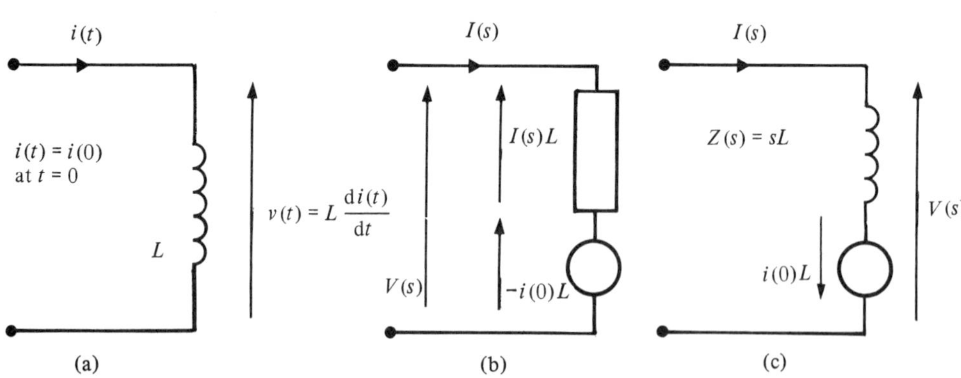

Fig. 1.8 The transformation of an inductor with current flowing at $t = 0$. a) The inductor in the time domain. b) The prototype representation of the inductor in the s domain. c) The impedance and initial condition form of the s domain circuit.

A similar procedure is shown for the transformation of a capacitor in Fig. 1.9. In this case, the initial condition is the capacitor voltage $v(0)$ and the defining expression is that for the current $i(t)$.

$$i(t) = C\frac{dv(t)}{dt} . \qquad \ldots(1.16)$$

Once again applying rule 11 from Table 1.1,

$$I(s) = C(sV(s) - v(0)) , \qquad \ldots(1.17)$$

$$= sCV(s) - Cv(0) . \qquad \ldots(1.18)$$

This expression is that for the sum of two currents in parallel. Drawing this in a component form requires an admittance sC and a current generator $-v(0)C$ as shown in Fig. 1.9b. This can be used directly, but a series equivalent is often more convenient. The conversion is achieved by application of Thevenin's theorem and the result is shown in Fig. 1.9c. The impedance is now $1/sC$ (compare with $1/j\omega C$) and the initial condition voltage generator, $\frac{v(0)}{s}$ is *in the same direction as the time domain initial voltage.* Note that, as with the inductor, the capacitor representation in the s domain *must* include both generator and impedance.

The transformation of resistors does not involve any differentials and thus there is no change: a resistor of R ohms in the time domain appears as an impedance of $Z(s) = R$ in the s domain.

At this stage, a first example can show how a simple network can be transformed into the s domain and analysed to find an output voltage, also in the s domain.

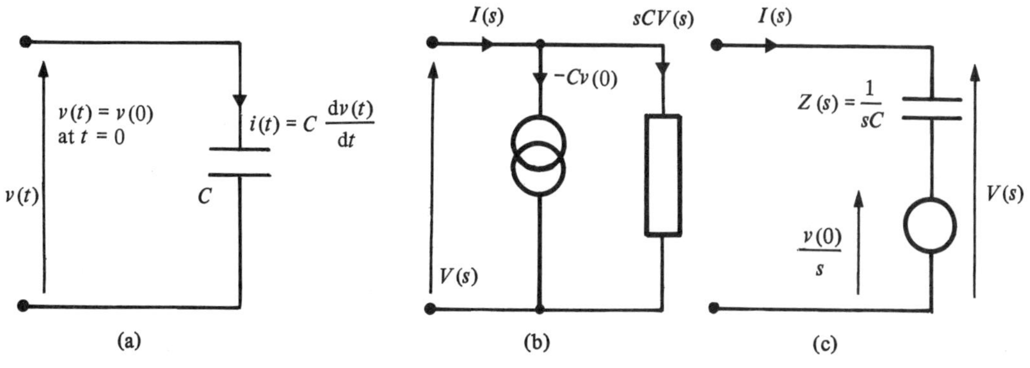

Fig. 1.9 The transformation of a capacitor with charge at $t = 0$. a) The capacitor in the time domain. b) The prototype circuit representation of the capacitor in the s domain. c) The impedance and initial condition form of the s domain circuit for the capacitor.

Example 1.1. A capacitor of 20 μF is fully charged from a 10 V d.c. supply and then connected across a circuit consisting of a 5 kΩ resistor in series with the parallel combination of a 1 μF capacitor and a 10 kΩ resistor. Draw the s domain circuit representing this situation and hence determine the voltage $V(s)$ that will appear across the 1 μF capacitor.

Solution. The time domain circuit is shown in Fig. 1.10a. The charged capacitor is shown with $v(0) = 10$ V and the connection at $t = 0$ is made by the switch SW. The 1 μF capacitor is uncharged (any earlier charge would have decayed through the 10 kΩ resistor), so $v(0)$ in this case is zero.

The transformed s domain circuit is shown in Fig. 1.10b. The switch is no longer present as this circuit only exists for t greater than zero when the switch is a short circuit. The resistors are unchanged and each capacitor is replaced by the equivalent shown in Fig. 1.9c. Note particularly the direction of the initial condition generators and that the generator for the 1 μF capacitor is present even though it has zero value.

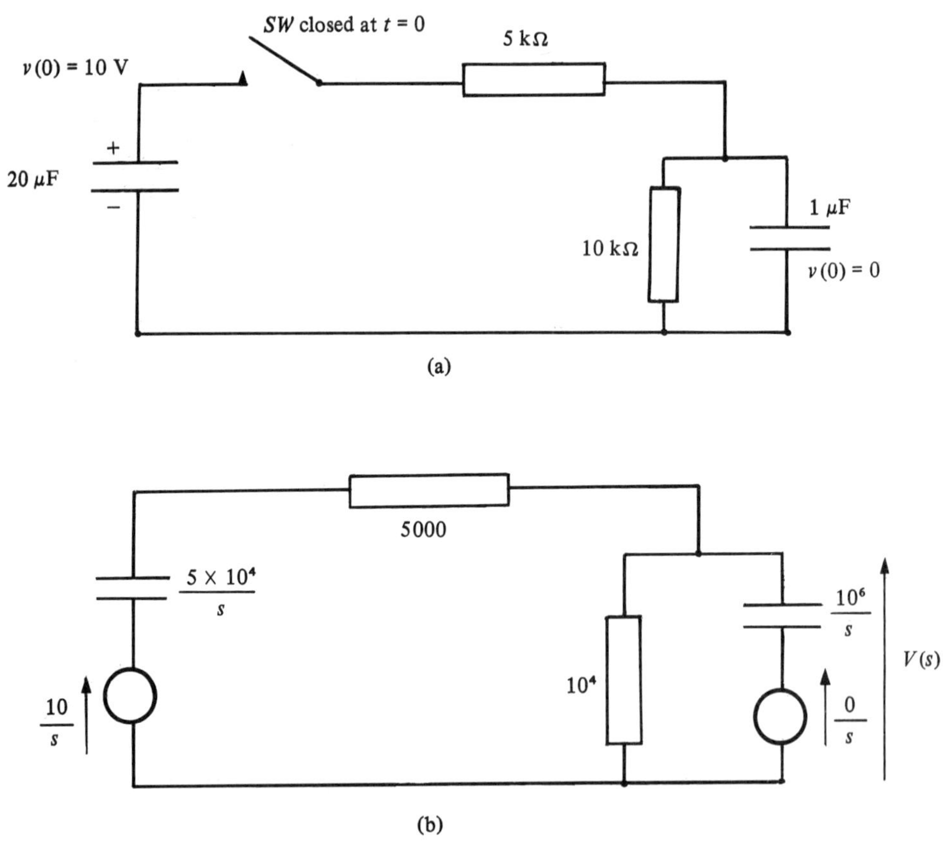

Fig. 1.10 Circuit for Example 1.1. a) The time domain circuit. b) The s domain circuit.

$V(s)$ may now be determined using d.c. methods. Kirchhoff's loop equations could be used but in this case (as $v(0) = 0$ for the 1 μF capacitor), it is easier to calculate the impedance $Z(s)$ for the parallel combination and then to find $V(s)$ by potential division. For the parallel combination,

$$Z(s) = \frac{10^4 \times 10^6/s}{10^4 + 10^6/s} \text{ ohms.} \qquad \ldots(1.19)$$

Eliminating the fractions in both numerator and denominator,

$$Z(s) = \frac{10^{10}}{10^4 s + 10^6} \text{ ohms.} \qquad \ldots(1.20)$$

Dividing through by 10^4,

$$Z(s) = \frac{10^6}{s + 100} \text{ ohms.} \qquad \ldots(1.21)$$

By potential division,

$$V(s) = \frac{10/s \times \dfrac{10^6}{s + 100}}{\dfrac{5 \times 10^4}{s} + \dfrac{10^6}{s + 100} + 5000} . \qquad \ldots(1.22)$$

Eliminating fractions by multiplying numerator and denominator by $s(s + 100)$,

$$V(s) = \frac{10^7}{5 \times 10^4 (s + 100) + 10^6 s + 5000s\,(s + 100)} . \qquad \ldots(1.23)$$

Collecting terms in s^2 and s, and dividing through by 5000;

$$V(s) = \frac{2000}{s^2 + 310s + 1000} . \qquad \ldots(1.24)$$

This is the required solution at this stage. Later in the chapter, we see how this result can be transformed back to the time domain to give the output voltage $v(t)$. Hence, the simple process of transformation, requiring some intermediate algebra, eliminates the need to solve the differential equation which describes the network in the time domain (see Figs. 1.3 and 1.4).

It is useful to stress the algebraic steps that have been used in the above solution – eliminate fractions in both numerator and denominator by multiplying both by suitable factors; collecting terms in powers of s; dividing through by the coefficient of the highest power of s. These steps will appear in almost every analysis.

1.8 TRANSFORMATION OF SIGNALS

The transformation of signals is achieved simply by reference to the Laplace transforms in Table 1.1. Some common situations are illustrated in the next example.

Example 1.2. The waveforms shown in Fig. 1.11 are to be applied to the network shown in Fig. 1.10a (instead of the 20 μF capacitor and switch). In each case, determine the value of the transformed signal and also the value of the initial condition generator associated with the 1 μF capacitor.

Solution. In each case, we are concerned with the transform for $t > 0$. *The initial conditions, however, are determined for the circuit conditions for* $t < 0$.

i. Since the voltage is zero before $t = 0$, there are no initial conditions. For $t > 0$, the signal is a constant 1 V; this may be written:

$$v(t) = u(t)\,,$$

a unit step.
Referring to Table 1.1,

$$V(s) = 1/s. \qquad \ldots (1.25)$$

ii. When t is less than zero, the applied voltage is 1 V. This is divided between the two resistors, making:

$$V_c = v_c(0) = 1 \times \frac{10}{15} = 0.67 \text{ volts.}$$

The i.c. generator is thus $+0.67/s$. For $t > 0$, the applied signal is –2 V, or $v(t) = -2u(t)$.

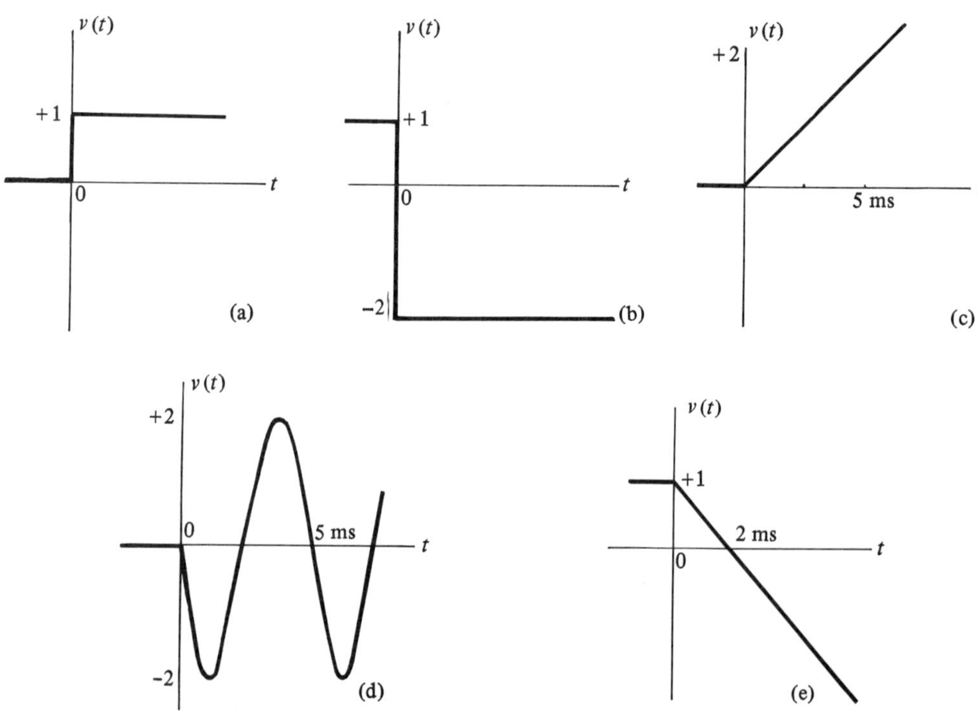

Fig. 1.11 The waveforms for the signals to be transformed in Example 1.2.

From Table 1.1:

$$V(s) = -2/s. \qquad \ldots(1.26)$$

Note: although the step is apparently of three volts' amplitude, from +1 to –2 V, it is the conditions obtaining for $t > 0$ that define $V(s)$.

iii. No initial conditions in this case and a ramp of $\frac{2}{5 \times 10^{-3}} = 400\ Vs^{-1}$ switched on at $t = 0$, with $v(t) = 0$. The expression for $v(t)$ is given by:

$$v(t) = 400\ tu(t);$$

$$\therefore \quad V(s) = \frac{400}{s^2}. \qquad \ldots(1.27)$$

Note: the $u(t)$ is required in the expression for $v(t)$ since the ramp starts at $t = 0$; without it, the ramp would continue for negative values of t.

iv. No initial conditions and a sinusoid switched on at the beginning of the negative half-cycle at $t = 0$. Since the period is 5 ms, the angular frequency $= \frac{2\pi}{5 \times 10^{-3}} = 1257$ rads^{-1}. The amplitude is 2 V.
From the table,

$$V(s) = \frac{2 \times 1257}{s^2 + (1257)^2}. \qquad \ldots(1.28)$$

v. The initial conditions are the same as in (ii), a generator of +0.67/s. The signal in this case includes a d.c. level of +1 V. and a negative ramp at 500 Vs^{-1}.

$$\therefore \quad V(s) = \frac{1}{s} - \frac{500}{s^2}. \qquad \ldots(1.29)$$

1.9 ANALYSIS IN THE s DOMAIN

Any of the techniques of network analysis may be employed in s domain analysis. With simple arrangements, the series-parallel relationships based on Ohm's law are probably the most convenient. More complicated circuits involving two or more loops can be solved using mesh or nodal analysis, but in some cases, reduction by Thevenin's or Norton's theorems may simplify the work. In some situations, $T \rightarrow \pi$ (Star → Delta) conversion for four-terminal network techniques can be advantageous. A variety of these methods are used in this chapter and throughout the remainder of the book.

1.10 TRANSFORMATION OF THE RESPONSE TO THE TIME DOMAIN

Functions of s can be transformed directly using integral calculus but in practice, it is easier to arrange functions so that they can be seen to fit one or more of the terms found in the table of transforms. This can be demonstrated using the response found for Example 1.1 in expression (1.24). Rewriting this result:

$$V(s) = \frac{2000}{s^2 + 310s + 1000} \quad \ldots (1.24)$$

Examination of the table shows no quadratic expressions of this form, but as a denominator term of $(s + a)$ appears in the transform for e^{-at}, this is the form to aim at.

Factorize the denominator using the formula:

$$s = \frac{-b \pm \sqrt{b^2 - 4ac}}{2a} = -3.26, \text{ or } -307. \quad \ldots (1.30)$$

The factors are therefore $(s + 3.26)$ and $(s + 307)$. Expression (1.24) can now be divided into partial fractions.

$$\frac{2000}{(s + 3.26)(s + 307)} = \frac{A}{s + 3.26} + \frac{B}{s + 307}. \quad \ldots (1.31)$$

Cross-multiply,

$$2000 = A(s + 307) + B(s + 3.26). \quad \ldots (1.32)$$

Equating coefficients of s;

$$A + B = 0, \quad \text{or} \quad A = -B, \quad \ldots (1.33)$$

and

$$2000 = 307A + 3.26B. \quad \ldots (1.34)$$

Substitute for B from (1.33) to find

$$A = 6.58 \text{ and } B = -6.58. \quad \ldots (1.35)$$

Hence,

$$V(s) = \frac{6.58}{s + 3.26} - \frac{6.58}{s + 307}. \quad \ldots (1.36)$$

Note: A and B could have been found more quickly by the 'cover up' rule (see Appendix A1).

$$A = \left.\frac{2000}{s + 307}\right|_{s = -3.26} = 6.58, \quad \ldots (1.37)$$

$$B = \left.\frac{2000}{s + 3.26}\right|_{s = -307} = -6.58. \quad \ldots (1.38)$$

This method is used wherever it is applicable for the remainder of this chapter. Finally, referring to (1.36) and to the Laplace transform tables,

$$v(t) = 6.58e^{-3.26t} - 6.58e^{-307t} \quad \ldots (1.39)$$

The following section of this chapter examines a number of examples illustrating the techniques described above and demonstrating sketching techniques for the time response.

1.11 INDUCTIVE CIRCUIT WITH SWITCHED d.c.

Example 1.3. The switch in the network shown in Fig. 1.12 is at position A for a long time and then moved to position B at $t = 0$. Calculate an expression for the resulting current in the 0.1 H inductor. By reference to a sketch of this time response, estimate the maximum value of this current.

Solution. From the original circuit, the initial inductor currents are zero for the 0.1 H inductor and $\frac{10}{20+5} \times \frac{1}{2} = 0.2$ A for the 0.2 H inductor (since $z = 0$, or a short-circuit, in the steady-state case for d.c. excitation where $\omega = 0$ and $z_L = \mathrm{j}\omega L$). The generator for the 0.2 H is therefore $0.2 \times 0.2 = 0.04$ V downwards (the direction of the original current). This, and the transformed components, are shown in Fig. 1.12b. Since the current in the 0.1 H inductor is required, the right-hand mesh can be reduced by Thevenin to give the simple series circuit shown in Fig. 1.12c.

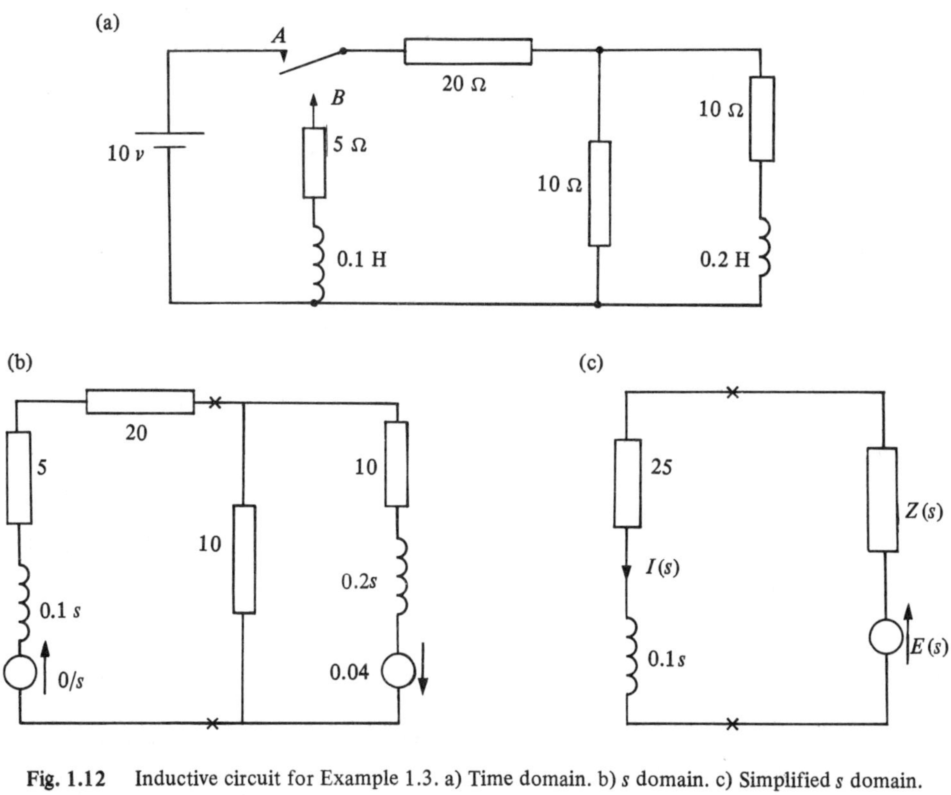

Fig. 1.12 Inductive circuit for Example 1.3. a) Time domain. b) s domain. c) Simplified s domain.

$$Z(s) = \frac{10\,(10 + 0.2s)}{10 + 10 + 0.2s} = \frac{10\,(s + 50)}{(s + 100)}, \qquad \ldots (1.40)$$

$$E(s) = \frac{-0.04 \times 10}{10 + 10 + 0.2s} = \frac{-2}{(s + 100)} \qquad \ldots (1.41)$$

The required $I(s)$ is now given by:

$$I(s) = \frac{\dfrac{-2}{(s + 100)}}{\dfrac{10\,(s + 50)}{(s + 100)} + 25 + 0.1s}, \qquad \ldots (1.42)$$

$$= \frac{-2}{10(s + 50) + 0.1\,(s + 250)\,(s + 100)}.$$

Multiplying out and collecting terms,

$$I(s) = \frac{-20}{s^2 + 450s + 30\,000}.$$

Factorizing,

$$I(s) = \frac{-20}{(s + 81.4)\,(s + 369)} = \frac{A}{(s + 81.4)} + \frac{B}{(s + 369)}. \qquad \ldots (1.43)$$

By cover up,

$$A = -0.065, \text{ and } B = +0.065.$$

From Table 1.1,

$$i(t) = -65\,(e^{-81.4t} - e^{-369t}) \text{ milliamperes.} \qquad \ldots (1.44)$$

To sketch this response, it is convenient to express the exponents in time constant form:

$$i(t) = (65e^{\frac{-t}{2.6}} - 65e^{\frac{-t}{12}}) \text{ milliamperes,}$$

where t is expressed in milliseconds.

The two components can now be sketched as shown in Fig. 1.13 and then added to give the total. The detailed procedure is as follows:

i. The time scale is chosen to allow at least twice the longest time constant. In this sketch, a scale up to 36 ms has been chosen.
ii. The vertical scale is chosen to allow for the maximum value of each term; in this case ±70 mA is sufficient.
iii. Each time constant is marked on the time axis together with further intervals equal to the time constants.
iv. Points are obtained for each of the terms at the time constant intervals by dividing the initial value (65) by e, e^2, and e^3.
v. If required, the slope at the start and at each point is such as to reach zero after a further time constant if continued as a straight line.

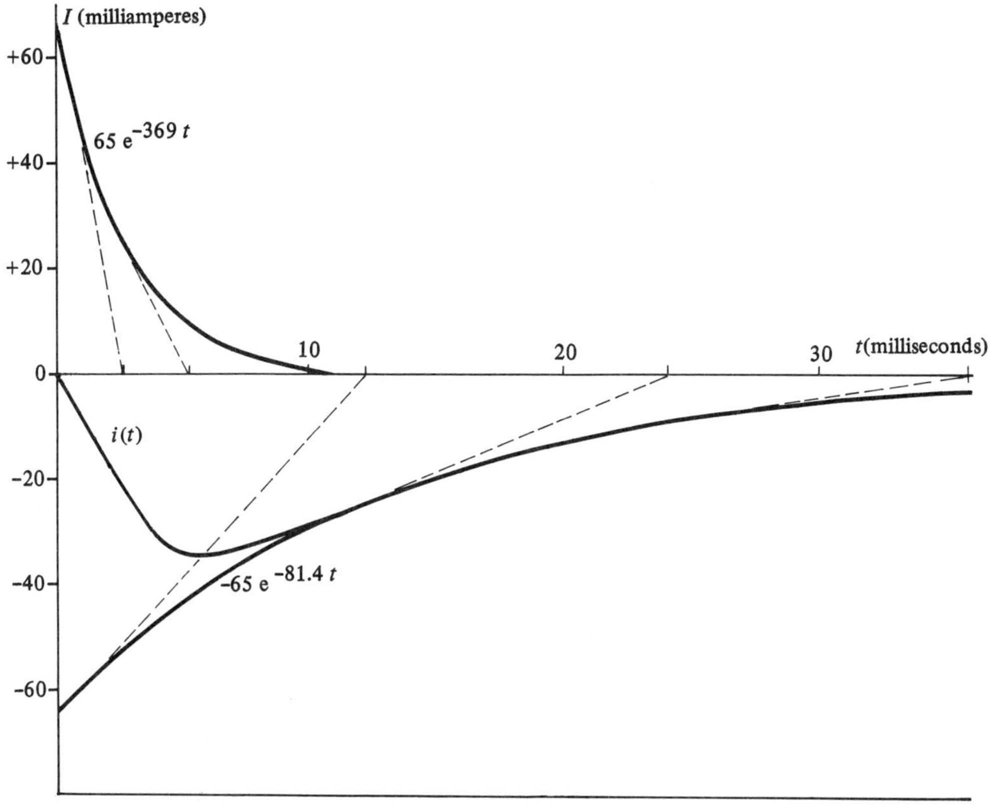

Fig. 1.13 The transient response for the solution of Example 1.3.

vi. Sketch in the exponent terms.
vii. Add at convenient points and sketch total as shown.

To complete the required solution, the estimated maximum current is –34 mA at about 5 ms.

1.12 CAPACITIVE CIRCUIT WITH SWITCHED d.c.

Example 1.4. In the circuit shown in Fig. 1.14, the switch is connected to position *A* for a long time and then moved to *B*. Calculate an expression for the voltage across the 0.2 μF capacitor during the time following this change. For this example, only an outline solution is given, the intermediate steps being left to the reader.

Solution. The positive 5 V, prior to switching, will charge both capacitors to 3 V. The transformed circuit will include a signal voltage of $\frac{-10}{s}$. This, together with the 2 kΩ and 3 kΩ resistors can be reduced by Thevenin's theorem to the circuit shown in Fig. 1.14b.

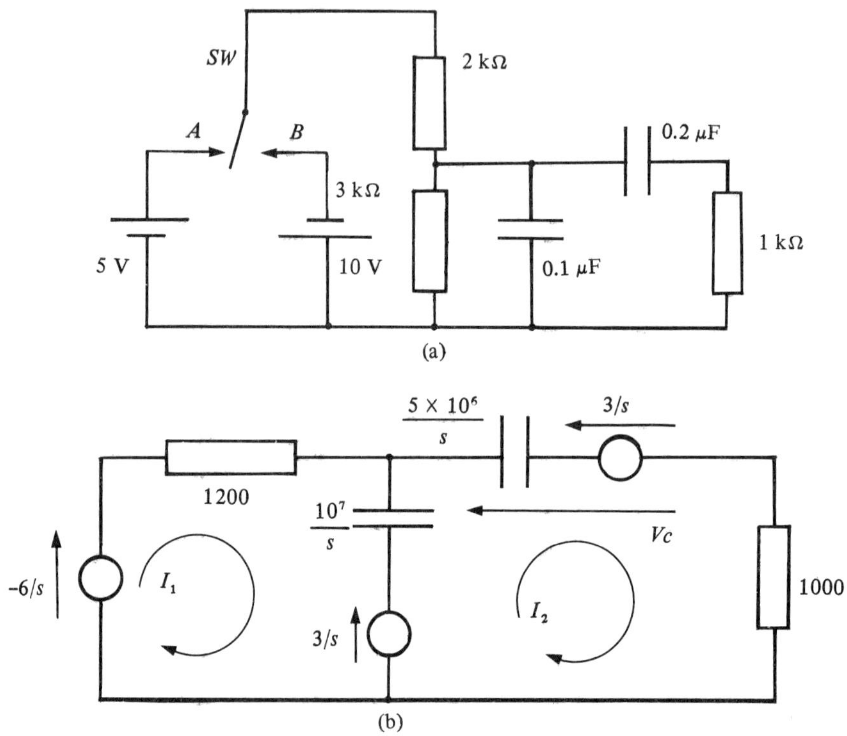

Fig. 1.14 The *RC* network for Example 1.4. a) Time domain. b) Simplified *s* domain.

Equations can then be written for the two meshes shown, and after simplification these are:

$$-9 = (1200s + 10^7)I_1 - 10^7 I_2, \qquad \ldots (1.45)$$

$$0 = -10^7 I_1 + (1000s + 1.5 \times 10^7)I_2. \qquad \ldots (1.46)$$

Solving by determinants and simplifying,

$$I_2 = \frac{-75}{(s + 1948)(s + 21\,400)}. \qquad \ldots (1.47)$$

The required V_C includes both the voltage drop across the capacitive impedance *and the $3/s$ generator.*

$$V_C = \frac{3}{s} - \frac{75 \times 5 \times 10^6}{s(s + 1948)(s + 21\,400)}. \qquad \ldots (1.48)$$

Dividing the second term into partial fractions, adding the first term and finding the inverse transform leads to the following result.

$$V_C = -6 + 9.9e^{-t/513} - 0.9e^{-t/47}, \qquad \ldots (1.49)$$

where t is expressed in microseconds.

1.13 SWITCHED RAMP APPLIED TO A *CR* CIRCUIT

Example 1.5. The waveform shown in Fig. 1.15a is applied to the scaled circuit shown in Fig. 1.15b. By use of Laplace transform techniques, determine the waveform for v_o. Illustrate the answer by means of a sketch.

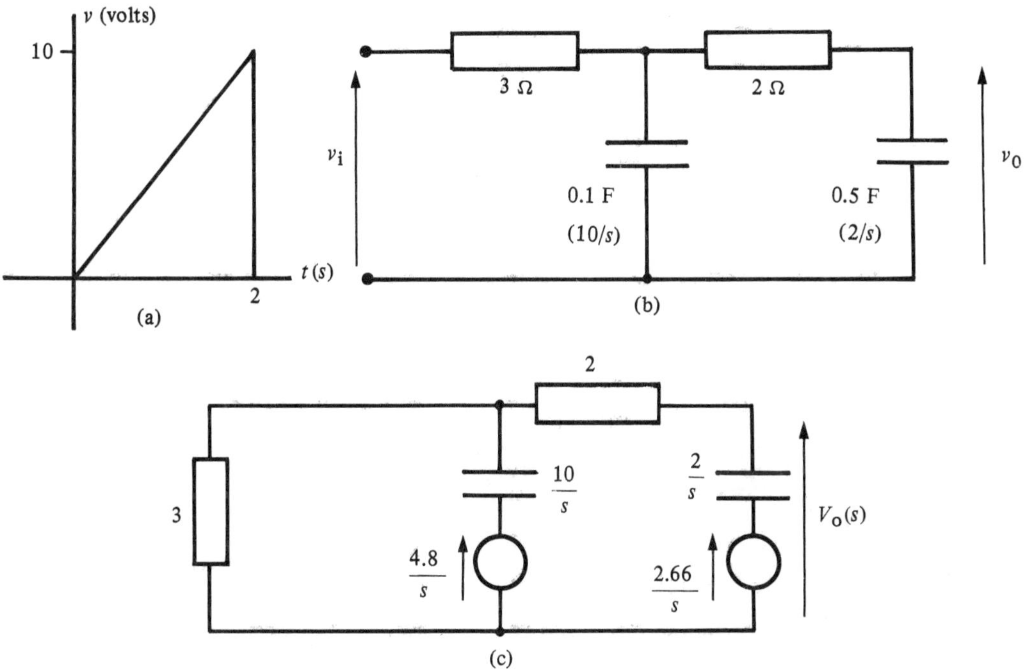

Fig. 1.15 Waveform and circuit for Example 1.5. a) Applied signal. b) Circuit for the first time period. c) The circuit for the second time period.

Solution. There are no initial conditions for the time period 0 to 2*s* and the applied signal is $5t$, which transforms into $\frac{5}{s^2}$. The remainder of the circuit is transformed as shown. (For further information regarding scaled or normalized circuit components, see Appendix A3.) Application of mesh analysis or Thevenin's theorem shows the current I in the 0.5 F capacitor to be given by:

$$I = \frac{8.33}{s\,(s + 0.37)\,(s + 8.96)}\,, \qquad \ldots(1.50)$$

and since $V_c = \frac{I}{sC}$,

$$V_c = \frac{16.67}{s^2\,(s + 0.37)\,(s + 8.96)}\,. \qquad \ldots(1.51)$$

Using partial fraction expansion,

$$= \frac{A}{s^2} + \frac{B}{s} + \frac{C}{(s+0.37)} + \frac{D}{(s+8.96)} \quad \ldots (1.52)$$

Note that the repeated root s^2 requires the extra term B/s. The cover up rule shows A, C and D to be 5, 14.17 and –0.024 respectively. Cross-multiplying now gives:

$$16.67 = 5(s+0.37)(s+8.96) + Bs(s+0.37)(s+8.96) + Cs^2(s+8.96) + Ds^2(s+0.37) \quad \ldots (1.53)$$

Equating the coefficients of s^2,

$$0 = B + C + D,$$

and

$$B = -C - D = -14.1 \quad \ldots (1.54)$$

Hence, from (1.53) and the tables,

$$v_c = 5t - 14.1 + 14.17e^{-0.37t} - 0.024e^{-8.96t} \quad \ldots (1.55)$$

This solution is only applicable up to $t = 2$; the final conditions at $t = 2$ provide the initial conditions for the second time period, $t > 2$. During this period, the signal is zero, but the capacitor initial charge must be determined from result (1.55). Substituting for $t = 2$ in (1.55) gives $v_c = 2.66$ V for the 0.5 F capacitor. For the 0.1 F capacitor, the voltage can be found from $2.66 + 2i$, where i is obtained by transforming (1.50). This calculation is left for the reader and results in a v_c of 4.8 V. The new circuit for the remaining time period is shown in Fig. 1.15c.

Analysis by Thevenin's theorem or mesh analysis results in:

$$v_c = 3.03e^{-0.37t} - 0.36e^{-8.96t} \quad \ldots (1.56)$$

where the t values are those for the second time period.

The sketch of the complete output waveform is shown in Fig. 1.16; the following steps have been used in its construction.

i. Determine the time constants: $\frac{1}{0.37} = 2.7$ s, $\frac{1}{8.96} = 0.11$ s.

ii. Choose a time scale: two seconds is specified for the first time period; for the second, allow twice the longest time constant or five seconds giving seven seconds total.

iii. Choose the vertical scale: the first two terms combined provide a ramp starting from –14.1 V, the third is an exponent starting from +14.17 and the fourth is negligible; a scale of ±15 V will therefore be sufficient.

iv. Draw $5t - 14.1$, a straight line from –14.1 at $t = 0$, with a slope of 5 Vs^{-1} to $t = 2$.

v. Draw $14.17e^{-t/2.7}$ using points for $t = 0$, 1.35 and 2.7. Ignore $0.024e^{-t/0.11}$ as it will only produce a slight delay during the first 0.2 seconds.

vi. Add (iv) and (v) at two or three convenient points.

vii. Sketch $3.03e^{-t/2.7}$ for the second time period.

viii. Note that at $t = 0$ for this time period, the total is approximately 2.66, which is the

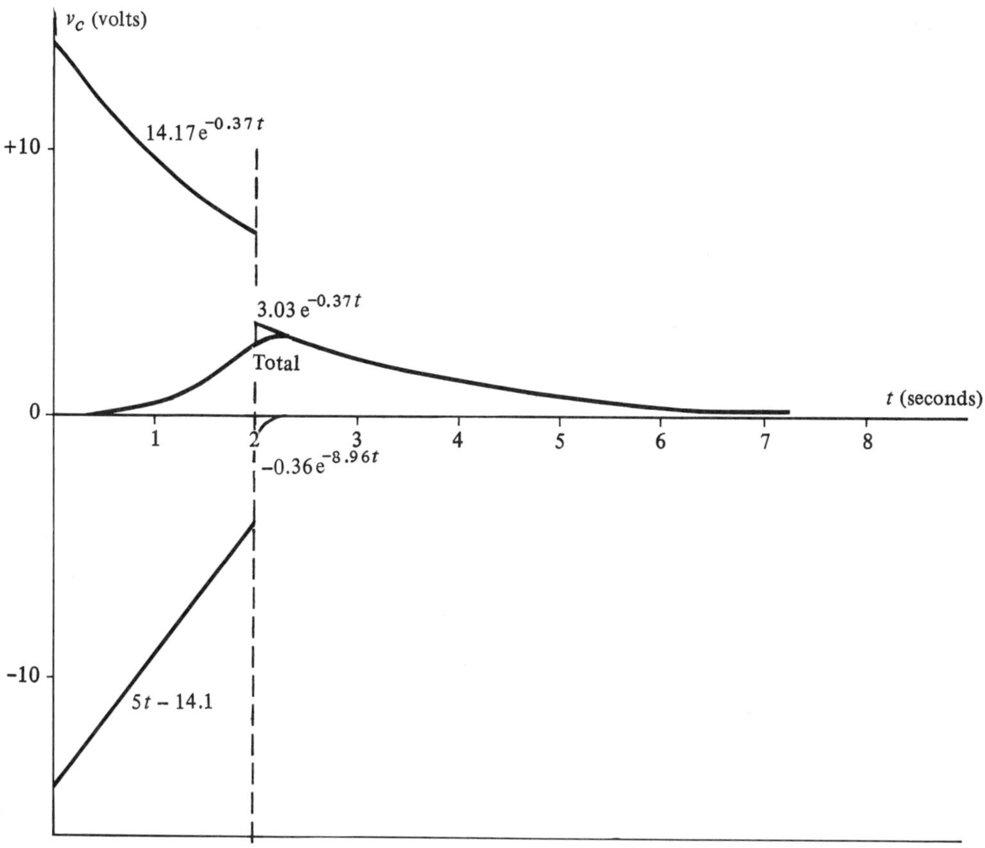

Fig. 1.16 The response waveform for the solution of Example 1.5.

same as the total for the first period at $t = 2$. (C cannot change charge instantaneously.)

ix. The term $0.36e^{-t/0.11}$ rounds off the top of the waveform as shown. It rises slightly as the 0.1 F continues to discharge into the 0.5 F after which they both discharge through the 3 Ω resistor.

1.14 SWITCHED SINUSOID APPLIED TO A *CR* CIRCUIT

Example 1.6. Fig. 1.17 shows a section of a simple, triac control circuit. The 50 Hz, 240 V_{rms} mains is applied to the load R_L in series with the control circuit. It may be assumed that the triac will be conducting during part of the positive half-cycle and ceases to conduct at the beginning of the negative half-cycle. Determine an expression for the voltage across the capacitor and, by means of a sketch, estimate the firing angle in the next half-cycle if the triac fires at a negative voltage of 100 V. Neglect the effect of R_L.

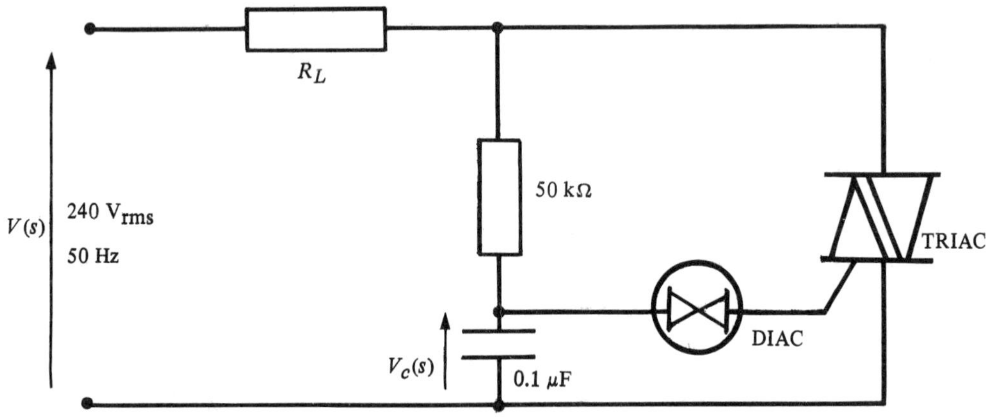

Fig. 1.17 The triac power control circuit for Example 1.6.

Solution. The transform for the applied voltage switched on at the beginning of the negative half-cycle is given by:

$$V(s) = \frac{-314 \times 240 \times \sqrt{2}}{s^2 + 314^2}. \qquad \ldots (1.57)$$

There are no initial conditions as C will have discharged while the triac is conducting. The control voltage V_c is obtained by potential division:

$$V_c(s) = V(s) \times \frac{1/sC}{R + 1/sC} = V(s) \times \frac{200}{s + 200}. \qquad \ldots (1.58)$$

Substituting for $V(s)$,

$$V_c(s) = \frac{-2.14 \times 10^7}{(s^2 + 314^2)(s + 200)}. \qquad \ldots (1.59)$$

Expanding into partial fractions,

$$V_c(s) = \frac{As + B}{s^2 + 314^2} + \frac{C}{s + 200}. \qquad \ldots (1.60)$$

The cover-up rule is then used to find C, after which, cross-multiplying and equating coefficients will provide A and B. Substituting in (1.60);

$$V_c(s) = \frac{154s - 30\,772}{s^2 + 314^2} - \frac{154}{s + 200}. \qquad \ldots (1.61)$$

The first term divides into $\frac{154s}{s^2 + 314^2}$ which transforms into a cosine term and into $\frac{-30\,772}{s^2 + 314^2}$ or $\frac{-98 \times 314}{s^2 + 314^2}$ which transforms into a sine term. Thus,

$$v_c(t) = 154 \cos 314t - 98 \sin 314t - 154e^{-200t}. \qquad \ldots (1.62)$$

This may be checked, as at $t = 0$,

$$v_c(t) = 154 - 0 - 154 = 0.$$

To sketch this waveform, it is convenient to express the sinusoidal terms in the form:

$$K \cos(\theta + \omega t) = K[\cos\theta \cos\omega t - \sin\theta \sin\omega t]; \quad \ldots(1.63)$$

from which,

$$\theta = \tan^{-1}\frac{98}{154} = 32°, \quad \ldots(1.64)$$

$$K = \sqrt{98^2 + 154^2} = 182. \quad \ldots(1.65)$$

Thus

$$v_c(t) = 182 \cos(314t + 32°) - 154e^{-t/0.005} \quad \ldots(1.66)$$

The sketch is shown in Fig. 1.18 and has been constructed using the following steps:

i. The time scale must allow for at least one cycle of the sinusoid, i.e. $2\pi/\omega$ or 20 ms or twice the time constant, whichever is longer.
ii. The voltage scale requires at least ±182 V.
iii. Sketch a cosine of amplitude 182 V (ignoring the 32°) by marking the 0°, 90°, 180°, 270° and 360° points. If more accuracy is required, take 45° intervals, but it is not usually necessary.
iv. Calculate the time equivalent of $32° = \frac{32}{360} \times 20$ ms = 1.7 ms. Then for the v_c waveform, indicate $t = 0$ at a point 1.7 ms after the start of the cosine wave.

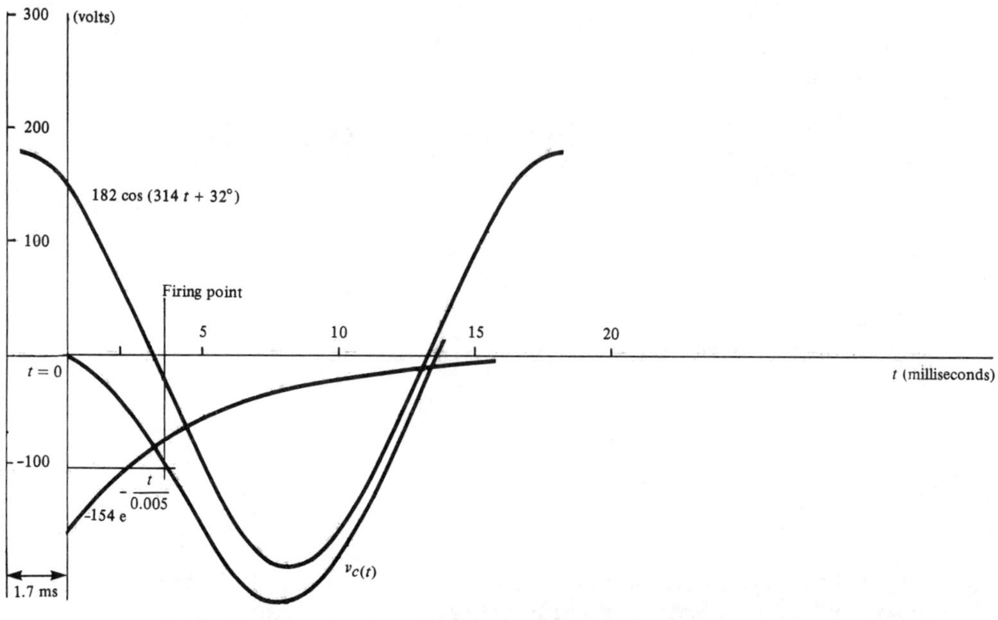

Fig. 1.18 The waveform for the control voltage in Example 1.6.

v. From point (iv), sketch $-154e^{-t/5}$ (t in ms) as shown.
vi. Add points as required to give the v_c waveform as shown.
vii. Estimate $v_c = -100$ V at $t = 3.4$ ms which represents a firing angle of $\frac{3.4}{20} \times 360 = 61°$.

1.15 SECOND ORDER CIRCUITS WITH COMPLEX CONJUGATE ROOTS

In several of the previous examples, quadratic factors with real roots have appeared in the denominator of the response expression. This has resulted in two exponentially decaying terms in the time domain. Alternative forms will have either two identical real roots or, more often, a pair of complex conjugate roots. Before proceeding to an example, it would be useful to consider the simple case shown in Fig. 1.19.

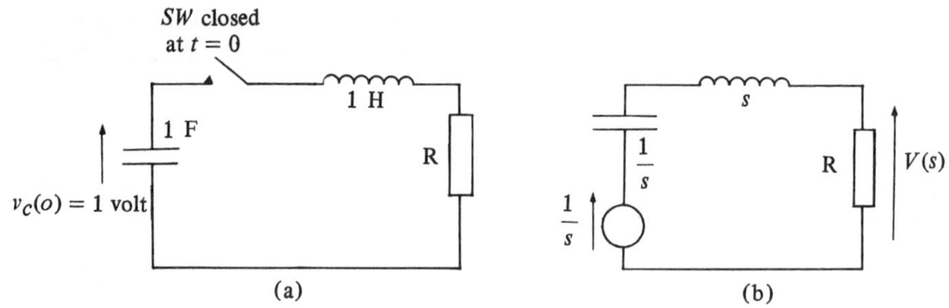

Fig. 1.19 A second order circuit which can have complex roots. a) Time domain. b) s domain.

From the transformed circuit,

$$V(s) = \frac{1}{s} \times \frac{R}{s + R + 1/s} = \frac{R}{s^2 + Rs + 1} . \qquad \ldots(1.67)$$

Solving the quadratic,

$$s = -\frac{R}{2} \pm \sqrt{\frac{R^2}{4} - 1} . \qquad \ldots(1.68)$$

If $R>2$, the roots are real and unequal.
If $R=2$, the roots are equal and expression (1.68) becomes:

$$V(s) = \frac{2}{(s+1)^2} . \qquad \ldots(1.69)$$

From Table 1.1, the time domain expression for this is given by:

$$v(t) = 2te^{-t} . \qquad \ldots(1.70)$$

The form of this function is shown in Fig. 1.20a. This curve is similar to those involving two exponentials, one positive and one negative. Note that no positive value of t will make the result negative.

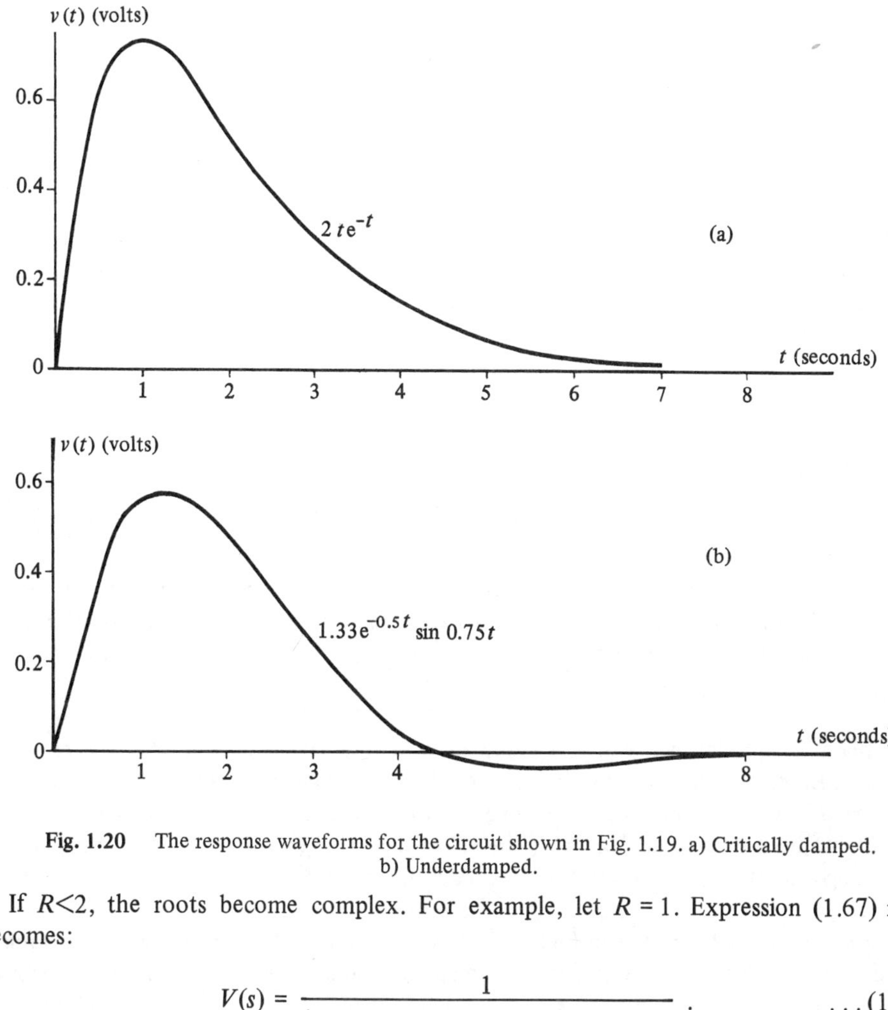

Fig. 1.20 The response waveforms for the circuit shown in Fig. 1.19. a) Critically damped. b) Underdamped.

If $R<2$, the roots become complex. For example, let $R = 1$. Expression (1.67) now becomes:

$$V(s) = \frac{1}{(s + 0.5 - j0.866)(s + 0.5 + j0.866)} \; . \qquad \ldots(1.71)$$

This can be divided into partial fractions from which a time domain result can be obtained. This approach is somewhat involved at this stage, but it is applied with the pole zero diagram in Chapter 3.

An alternative is to note from Table 1.1 that two of the transforms have denominators of the form: $(s + a)^2 + \omega^2$. The denominator for the case under consideration is $s^2 + s + 1$, which may be written in the required form by 'completing the square'.

Expanding the required form:

$$(s + a)^2 + \omega^2 = s^2 + 2as + a^2 + \omega^2 \; . \qquad \ldots(1.72)$$

Equating the coefficients of s with those in the original quadratic:

$$2a = 1, \text{ or } a = 0.5 \, ,$$

and

$$1 = a^2 + \omega^2 \text{ or } \omega^2 = 1 - 0.25 = 0.75 \ . \qquad \ldots(1.73)$$

Hence,

$$\omega = 0.87$$

$$V(s) = \frac{1}{(s+0.5)^2 + (0.87)^2} \qquad \ldots(1.74)$$

With reference to the table of transforms, this is the correct denominator for either $e^{-0.5t} \sin 0.75t$, or $e^{-0.5t} \cos 0.75t$. The second of these, however, requires $(s + a)$ in the numerator which for the sine form is simply ω.

Expression (1.75) can be rewritten:

$$V(s) = \frac{0.75 \times 1/0.75}{(s+0.5)^2 + (0.75)^2} \ . \qquad \ldots(1.75)$$

From which,

$$v(t) = 1.33e^{-0.5t} \sin 0.75t \ . \qquad \ldots(1.76)$$

The waveform for this decaying or damped sinusoid is shown in Fig. 1.20b.

This result is a special case and, in general, responses will include both sine and cosine terms which, after combination, produce an expression of the form $A\,e^{-\sigma t} \sin(\omega t + \theta)$. The next example illustrates this form and suggests a technique for sketching the result.

Example 1.7. A circuit consisting of inductance 5 H in series with resistance 100 Ω carries a direct current of 4 A. To avoid arcing when the supply is switched off, a 200 μF capacitor is connected across the circuit. (i) Calculate an expression for the voltage across the switch after it has been opened and (ii) with the aid of a sketch, estimate the maximum value of this voltage and the time at which this maximum occurs.

Solution. The circuit diagram in Fig. 1.21 shows the time and complex frequency domain circuits respectively. In the time domain circuit, we can see the d.c. supply from which the 4 A is obtained. This will have no effect upon the current $I(s)$ in the s domain circuit, but it must be included in the expression $V_{sw}(s)$ for the voltage across the open switch.

From Fig. 1.21a, with the switch closed and direct current flowing, there will be no voltage drop across the inductor and 4 × 100 or 400 V dropped across the resistor; both V_{dc} and $v(0)$ are therefore +400 in the directions shown. The inductor current $i(0)$ is 4 A as specified. In Fig. 1.21b, the initial condition generators are $i(0)L$ or 4 × 5 V and 400/s. Having chosen the sense of the unknown current $I(s)$, $V_{sw}(s)$ in the direction shown will be given by:

$$V_{sw}(s) = \frac{5000}{s} I(s) - \frac{400}{s} + \frac{400}{s} = \frac{5000}{s} I(s) \ . \qquad \ldots(1.77)$$

From Fig. 1.21b,

$$I(s) = \frac{20 + 400/s}{5s + 100 + \dfrac{5000}{s}} = \frac{4s + 80}{s^2 + 20s + 1000} \ . \qquad \ldots(1.78)$$

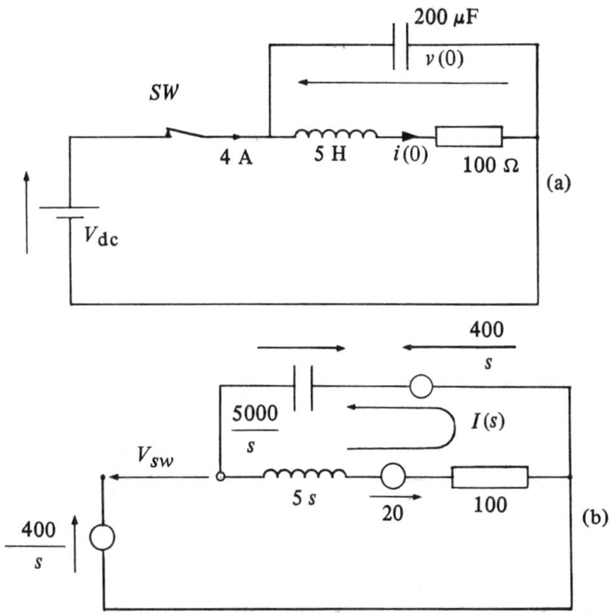

Fig. 1.21 The switched inductive circuit for Example 1.7. a) Time domain circuit. b) *s* domain circuit.

The quadratic has complex roots but completing the square should be left until partial fractions for the final answer have been taken.
From expression (1.77),

$$V_{SW}(s) = \frac{5000(4s + 80)}{s(s^2 + 20s + 1000)} = \frac{A}{s} + \frac{Bs + C}{s^2 + 20s + 1000}. \qquad \ldots (1.79)$$

By the cover-up rule,

$$A = \frac{5000 \times 80}{1000} = 400. \qquad \ldots (1.80)$$

Cross-multiplying;

$$5000\,(4s + 80) = 400\,(s^2 + 20s + 1000) + Bs^2 + Cs. \qquad \ldots (1.81)$$

Equating coefficients for s^2,

$$0 = 400 + B \quad \text{or} \quad B = -400,$$

and for *s*,

$$20\,000 = 8000 + C \quad \text{or} \quad C = 12\,000. \qquad \ldots (1.82)$$

Substituting in expression (1.80),

$$V_{SW}(s) = \frac{400}{s} + \frac{-400s + 12\,000}{s^2 + 20s + 1000}. \qquad \ldots (1.83)$$

Completing the square and putting the numerator in brackets:

$$V_{sw}(s) = \frac{400}{s} - \frac{(400s - 12\,000)}{(s+10)^2 + 30^2} \, . \qquad \ldots(1.84)$$

The numerator of (1.85) must now be arranged to match the form $K_1(s + a) \pm K_2\omega$.

$$V_{sw}(s) = \frac{400}{s} - \frac{400\,(s+10) - \dfrac{16\,000}{30} \times 30}{(s+10)^2 + 30^2} \, . \qquad \ldots(1.85)$$

Writing the inverse transform;

$$V_{sw} = 400 - (400e^{-10t} \cos 30t - 523e^{-10t} \sin 30t) \, . \qquad \ldots(1.86)$$

Applying the same simplification as that used in Example 1.6,

$$V_{sw} = 400 - 667e^{-10t} \cos(30t + 53.1°) \, . \qquad \ldots(1.87)$$

Note: the method used here is somewhat lengthy and a much shorter technique will be introduced in Chapter 3. This will, in fact, result in the sine form of $V_{sw} = 400 + 667e^{-10t} \sin(30t - 36.9°)$ which is the same as result (1.87).

The waveform for V_{sw} is shown in Fig. 1.22 together with the intermediate construction steps.

i. Calculate the period of the sinusoid T_p, the time corresponding to the phase angle T_ϕ and the time constant for the exponential decay τ. $T_p = \frac{2\pi}{\omega} = 0.21 \triangleq 0.2$ s, $T_\phi = 0.033 \triangleq 0.03$ s, $\tau = 0.1$ s. These figures are rounded off to make the sketching easier. This is justified as the sketch is only intended to give an approximate waveform.
ii. Choose the time scale to allow two periods for the cosine (0 to 0.4 s). Divide this into quarter periods and mark points corresponding to ±1 or 0. In this case, as we have –cos, these points are –1, 0, +1, 0, etc. Sketch the waveform for $-\cos 30t$.
iii. Since $-\cos(30t + 53.1°)$ is required, the time axis is shifted by T_ϕ to the right. With reference to this new axis, the waveform drawn for (ii) is now correct.
iv. Using the corrected time scale, sketch $667e^{-t/0.1}$. Note, this requires a different vertical scale.
v. (iii) and (iv) must be multiplied together: the points 0, –1 and +1 are sufficient; these are shown with a cross in the sketch.
vi. Calculate the starting point from $-667e^{\circ} \cos 53.1° = -400$ and sketch $-667e^{-t/0.1} \cos(30t + 53.1°)$.
vii. To obtain V_{sw}, 400 must be added to (vi). This is achieved by shifting the y axis by 400 V.

From the graph in Fig. 1.22, the required maximum voltage across the switch is about 740 V and it occurs at about 65 ms after opening the switch. If the reader repeats this example with a 100 μF capacitor, the corresponding voltage maximum exceeds 1000 V after about 35 ms.

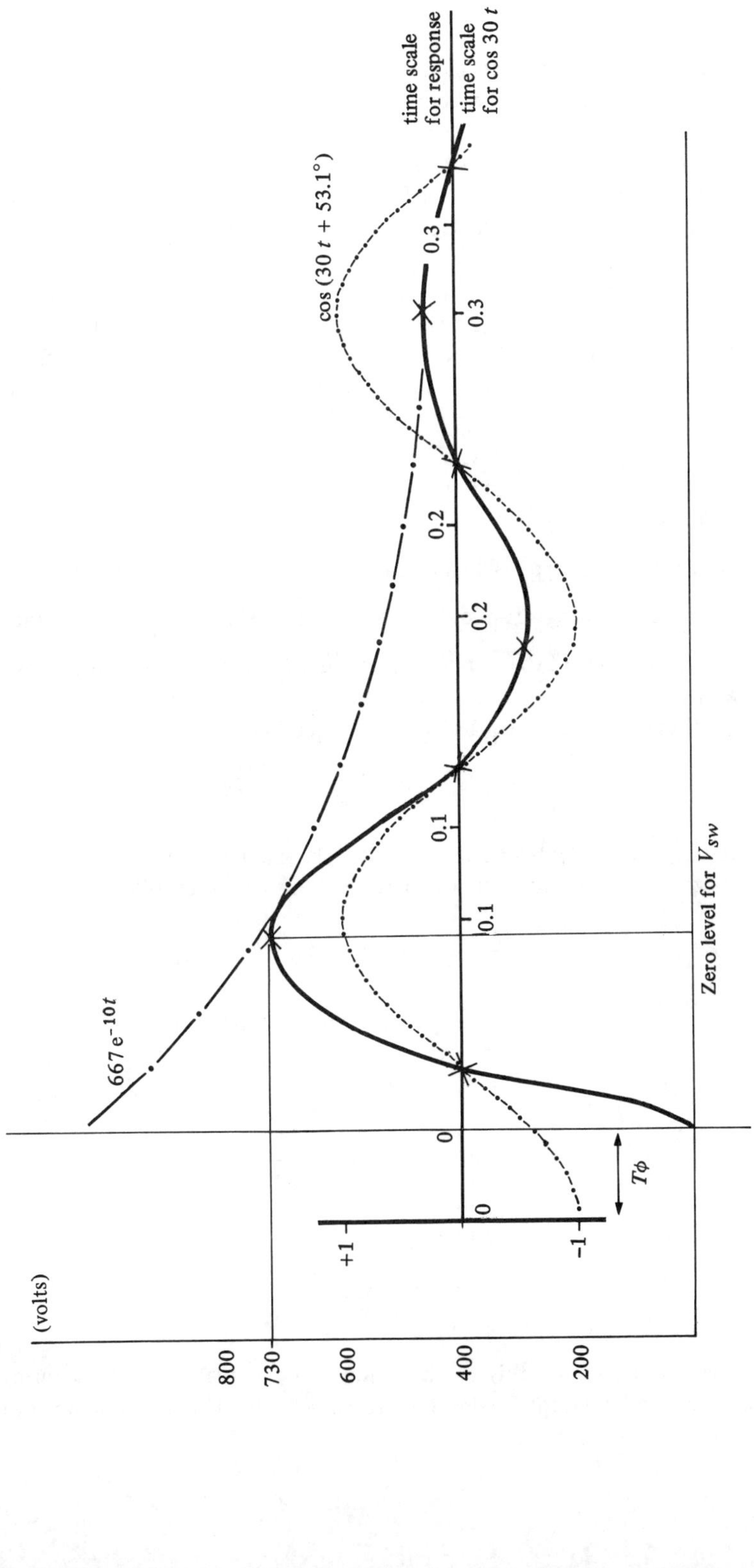

Fig. 1.22 The construction of the response waveform V_{sw} for Example 1.7.

1.16 ANALYSIS OF NON-ELECTRICAL SYSTEMS

With non-electrical systems, transformation into an equivalent s domain circuit is not always possible. The nearest equivalent is a block diagram with the transfer function of each block calculated separately. The determination of transfer functions for both electrical and some mechanical 'blocks' is described in Chapter 2. Examples involving block diagrams appear in other chapters. However, it will be useful to consider a simple mechanical example from basic principles, showing how the Laplace transform can be used in such cases.

Example 1.8. A rotary mechanism has an effective inertia of 0.2 kg/m^2. A steady torque of 5 Nm is suddenly applied to the mechanism at rest, but as it rotates, viscous friction produces a braking torque of 10^{-2} Nm/rad s^{-1}. Determine an expression for the angular velocity of the mechanism and hence find the final, steady velocity and the time elapsed before this condition arises.

Solution. When a torque is applied to an inertia load, the angular position θ of the load will change. θ accelerates from zero and the torque T can be equated to (the inertia, J) $\times$ (the angular acceleration, $\frac{d^2\theta}{dt^2}$). When θ is changing, there will be a braking torque due to the viscous friction given by (friction, F_v) $\times$ (angular velocity, $\frac{d\theta}{dt}$). This torque subtracts from the applied torque. If the applied torque is constant, the system settles to a final, steady velocity ω.

Expressing these relationships in a differential equation:

$$5u(t) - 10^{-2}\frac{d\theta}{dt} = 0.2\frac{d^2\theta}{dt^2} \; . \qquad \ldots(1.88)$$

Referring to the Laplace transforms in Table 1.1, and noting that the initial conditions (θ and ω) are zero at $t = 0$, equation (1.88) can be transformed as follows:

$$\frac{5}{s} - 10^{-2}s\theta(s) = 0.2s^2\theta(s) \; ; \qquad \ldots(1.89)$$

$$\frac{5}{s} = (0.2s^2 + 10^{-2}s)\,\theta(s) \; ;$$

from which,

$$\theta(s) = \frac{5}{s\,(0.2s^2 + 10^{-2}s)} \; . \qquad \ldots(1.90)$$

Rearrange to obtain the standard form,

$$\theta(s) = \frac{25}{s^2\,(s + 5 \times 10^{-2})} \; . \qquad \ldots(1.91)$$

The inverse transform for this result would show a final ramp representing a steady increase in θ, or in other words, a constant angular velocity. This problem, however, asks for an expression for the velocity ω.

In the s domain, $\dfrac{d\theta}{dt}$ transforms to $s\theta(s)$. Hence, velocity

$$s\theta(s) = \omega(s) = \frac{25 \times s}{s^2 (s + 5 \times 10^{-2})},$$

$$= \frac{25}{s(s + 5 \times 10^{-2})}. \qquad \ldots(1.92)$$

Standard inverse transform techniques result in:

$$\omega(t) = 500\,(1 - e^{-t/20}) \text{ radians per second}.$$

This may be converted to a steady velocity of 4775 rev/min after about one minute.

In this chapter, the reasons for using Laplace transform techniques have been introduced with special reference to electrical systems, followed by a brief examination of the meaning of complex frequency s. The procedures for the transformation of electrical circuits and signals were described in detail and the required techniques were illustrated for a wide range of circuits and signals. In many cases, these procedures are somewhat lengthy and one of the major objectives of the later chapters is to dispense with much of the algebraic manipulation by working directly from 'pole zero' diagrams. In some of the examples, the time domain response waveform was illustrated with a sketch. It is well worth mastering the technique of constructing such sketches as the result is often more meaningful than an expression for $f(t)$. It also serves as a useful check on the answer obtained, as an error may often be noticed with common sense. Finally, if an accurate numerical answer is required, the sketch may indicate a small time-range for which calculations should be made. The final section indicated the alternative approach that is required when Laplace methods are used for non-electrical systems.

1.17 EXAMPLES FOR FURTHER PRACTICE

Example 1.9. In the circuit shown in Fig. 1.23, switch SW is closed at $t = 0$. By the use of Laplace transform methods, calculate an expression for the current $i_2(t)$. By means of a sketch of the waveform of this current, estimate the maximum value and the time at which it will be half this maximum. ($0.4 - 0.35e^{-132t} - 0.05e^{-568t}$, 0.4 A at 4.5 ms.)

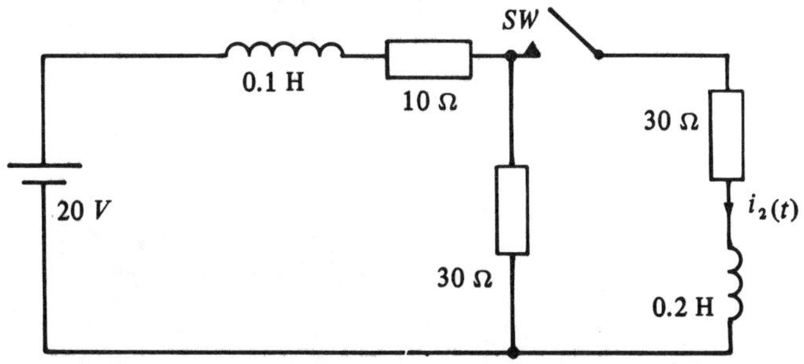

Fig. 1.23 Circuit for Example 1.9.

Example 1.10. With reference to the circuit shown in Fig. 1.24, determine an expression for the voltage $v_o(t)$, (i) if switch *SW* is opened after a long time, and (ii) if switch *SW* is then closed after a long time. ($-1.8e^{-100t}$, $0.978e^{-114t} + 11.022e^{-1169t}$.)

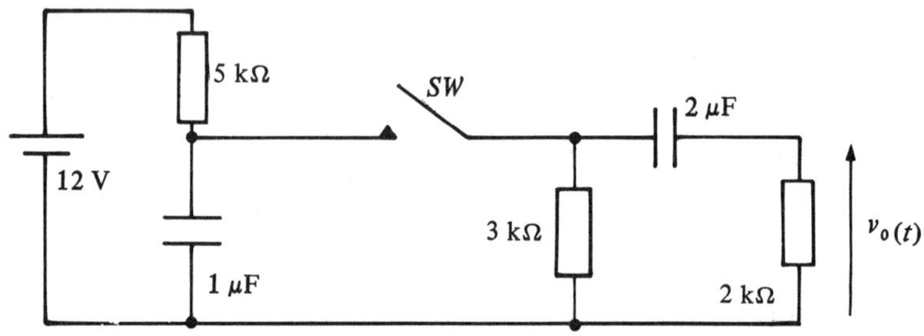

Fig. 1.24 Circuit for Example 1.10.

Example 1.11. Determine an expression for $v_o(t)$ in the circuit shown in Fig. 1.25 when the switch is moved from *A* to *B*, (i) if *R* is 4 Ω and (ii) if *R* is 50 Ω. For case (ii), sketch the variation of $v_o(t)$ with time and estimate the time of the first zero crossing and the maximum negative value. ($40.67e^{-10.1t} - 0.67e^{-617t}$, $52.7e^{-26t}\sin(66.5t + 69°)$, 29 ms – 14 V.)

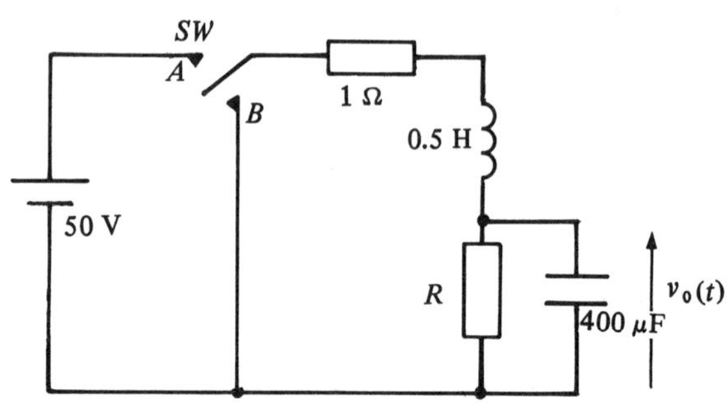

Fig. 1.25 Circuit for Example 1.11.

Example 1.12. In each of the circuits shown in Fig. 1.26, a 10 V step is applied at the terminals shown. Assuming that all the initial conditions are zero, calculate the indicated output in the time domain. ($11.7e^{-524t} - 1.7e^{-76t}$, $0.645(1 - e^{-590t})$.)

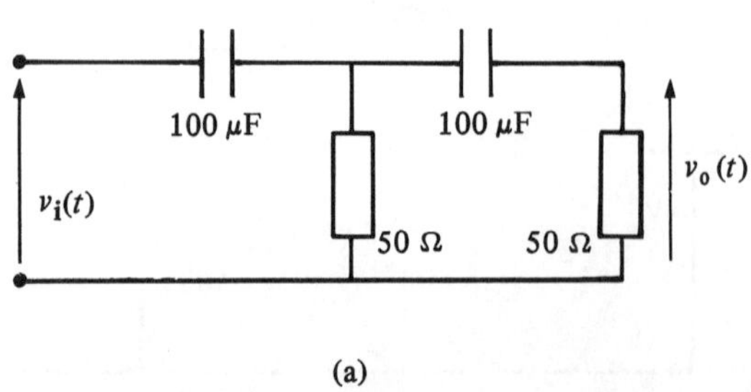

(a)

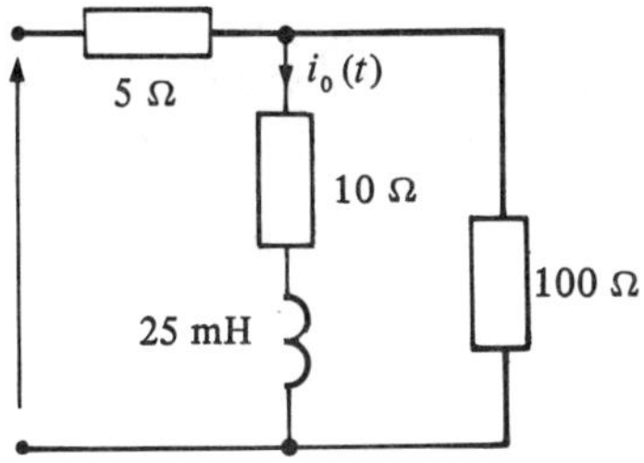

(b)

Fig. 1.26 Circuits for Example 1.12.

Example 1.13. If each of the three input signals listed below is separately applied to the compensator circuit shown in Fig. 1.27, calculate an expression for the output voltage shown. (i) A steady –2 V, switched to +3 V at $t = 0$. (ii) zero, with a ramp switched on at $t = 0$ passing through +20 V at 10 ms. (iii) A 500 Hz, 5 V peak sine wave switched on at $t = 0$ when $v = 0$. Illustrate each response with a sketch. ($3 - 4.05e^{-805t} + 4.05e^{-124\,200t}$, which = 0 at t = 11 μs and 370 μs; $2000t - 2.03 + 2.02e^{-805t} + 0.01e^{-124\,200t}$; $0.97e^{-805t} - 0.102e^{-124\,200t} + 1.48\sin(3140t - 36.1°)$, first maximum is 2 V at t = 0.65 ms, first zero at t = 1.3 ms.)

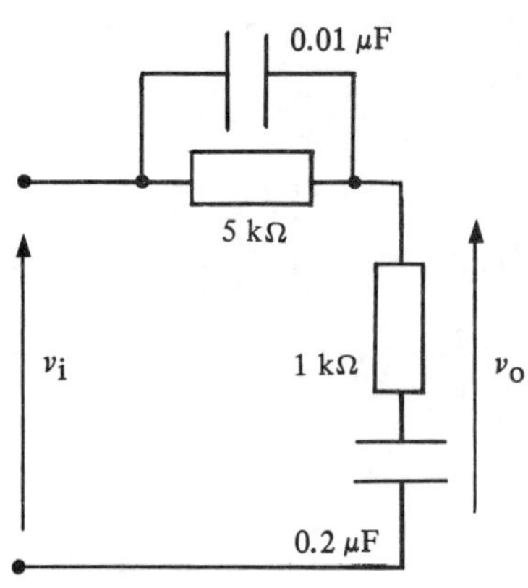

Fig. 1.27 Circuit for Example 1.13.

Example 1.14. A steady –2 V is applied to the circuit shown in Fig. 1.28 and then at $t = 0$, the input is switched to +3 V. Derive an expression for the output voltage $v_o(t)$. ($1.92 + 2.72e^{-14.6t} + 0.479e^{-21.4t}$, t in milliseconds.)

Example 1.15. A series *LCR* circuit has inductance 10 mH, capacitance 0.01 μF and resistance 1200 Ω. Derive expressions for the circuit current and capacitor voltage after a 10 V step has been applied to the combination. ($12.5e^{-60t}\sin 80t$ milliamperes, $10 - 12.5e^{-60t}\sin(80t + 53.1°)$ volts, t in milliseconds.)

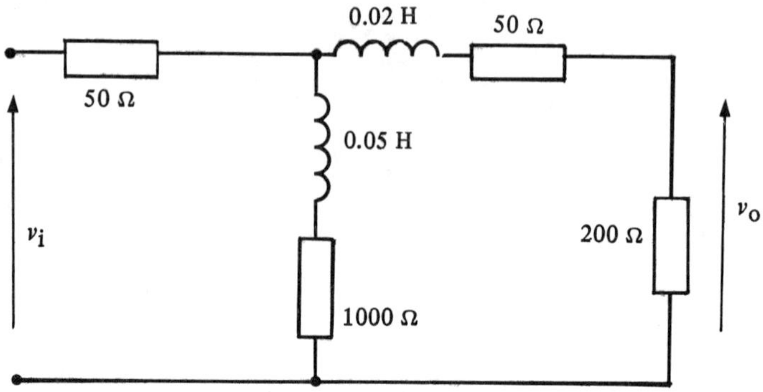

Fig. 1.28 Circuit for Example 1.14.

Example 1.16. A scaled d.c. circuit consists of a coil, of inductance 5 H and resistance 2 Ω, in parallel with a 0.5 F capacitor. The circuit is connected through a switch *SW* to a d.c. supply and when *SW* is closed, the resulting steady current is 2 A. If, at $t = 0$, *SW* is opened, draw the resulting circuit in the complex frequency (s) domain. Calculate an expression for the transient current $I(s)$ flowing in the circuit and hence determine the time response $i(t)$. Estimate the maximum voltage across the switch and the time at which it occurs. (7.3 V at 3 s.)

2

Transfer Functions

2.1 INTRODUCTION

The term transfer function is used to describe the external input–output properties of a system. Since these properties can be described in terms of the poles and zeros of the transfer function, this concept is essential to the theme of this book. The nature of a transfer function is quite independent of the usual system classification. Examples of biological, thermal, mechanical, electrical, chemical or other systems could all be described by the same form of transfer function. They would differ, in some cases dimensionally, or perhaps in the magnitude of certain constants, but the nature of the behaviour determined by the transfer function is common to all classes.

The application of transfer function methods is particularly useful in the fields of electrical engineering and control engineering, and in this chapter we examine the techniques for the determination of transfer functions for systems in these areas. Before this can be done, however, the various forms of transfer function must be defined together with their critical frequencies known as poles and zeros.

2.2 TRANSFER FUNCTION DEFINITIONS

The transfer function of a circuit or system is defined as the ratio of the output quantity to the input quantity expressed in terms of s with no initial conditions.

Thus, the transfer function $T(s)$ is given by:

$$T(s) = \frac{\text{Output}}{\text{Input}}(s)\ . \qquad \ldots(2.1)$$

Notice that the dimensions of output and input are not specified. In many cases, input and output have the same dimensions: e.g. $\frac{V_o}{V_i}$, $\frac{\theta_o}{\theta_i}$, etc. in which case $T(s)$ is dimensionless and may be conveniently regarded as a form of gain or amplification. Other forms are equally common; $\frac{I_o}{V_i}$ is a transfer admittance (Siemen) and $\frac{T_o}{V_i}$ expresses the output torque of a system in terms of the input voltage (Newton metre per volt).

The importance of a transfer function of a system or circuit is that it may be used to predict both the transient and the steady-state response resulting from a signal applied to the input. The transient response is usually expressed as a function of time and is the result of a switched input signal. The use of the transfer function for determining the transient response is discussed in Chapter 3. The steady-state response is the circuit output when a continuous signal is applied to the input and all transient terms are insignificant or as $t \to \infty$. Such continuous signals are usually sinusoids in which case the response is expressed as a function of frequency $j\omega$ (including d.c. where $\omega = 0$). Since $j\omega$ is a special case of the complex frequency s, the transfer function $T(s)$ becomes $T(j\omega)$ and provides direct information about the variation of response with signal frequency. The use of transfer functions for steady-state response is discussed in Chapter 4.

2.3 FORMS OF $T(s)$

Transfer functions are usually arranged into one of three forms, the first and most fundamental being the ratio of two polynomials in s.

$$T(s) = \frac{a_1 s^n + a_2 s^{n-1} + a_3 s^{n-2} + \dots a_n}{b_1 s^m + b_2 s^{m-1} + b_3 s^{m-2} + \dots b_m} \quad \dots (2.2)$$

This form is often modified to make the coefficients of the highest powers of s equal to unity.

$$T(s) = \frac{K\,(s^n + a_2/a_1 s^{n-1} + a_3/a_1 s^{n-2} \dots a_n/a_1)}{(s^m + b_2/b_1 s^{m-1} + b_3/b_1 s^{n-2} \dots b_m/b_1)} \quad \dots (2.3)$$

It is useful at this stage to note that for systems that are physically realizable, there are limitations on the form of the two polynomials. For example, in some cases, n cannot exceed m by more than one.

An alternative and more useful form is obtained by factorizing the polynomials in (2.3) to give:

$$T(s) = \frac{K(s+z_1)(s+z_2)\dots(s+z_n)}{(s+p_1)(s+p_2)\dots(s+p_m)} \quad \dots (2.4)$$

This form will be referred to as the pole zero or *PZ* form as it is the form from which the poles and zeros are obtained directly. For higher order polynomials, the factorization may cause a problem, but either numerical methods, or a technique using root locus plots (see

Chapter 6) can be used. Again, there will be restrictions on possible values of z_n and p_m. These will either be real, or in complex conjugate pairs. In the case of p_m, a further limitation is that the system will only be stable if the real values are positive or if the real parts of the complex conjugate pairs are positive.

The third form may be referred to as the Bode or time constant form and it can be obtained by rearranging expression (2.4).

$$T(s) = \frac{K_b\,(1 + T_1 s)\,(1 + T_2 s)\,...\,(1 + T_n s)}{(1 + T_a s)\,(1 + T_b s)\,...\,(1 + T_m s)} \, . \qquad ...(2.5)$$

This form is particularly useful for the determination of the steady-state frequency response discussed in Chapter 4. The values of T in this expression are the time constants of the system. The result must, however, be modified for quadratic factors with complex roots. This too, is discussed in Chapter 4.

2.4 POLES AND ZEROS

The poles and zeros of a transfer function are those *critical values* of complex frequency s which determine both transient and steady-state behaviour of the circuit or system. The zeros are the values of s which make $T(s)$ equal to zero. Referring to expression (2.4), we can see that the zeros of $T(s)$ are $-z_1$, $-z_2$, ... and $-z_n$. Similarly, the poles of $T(s)$ are the values of s which make $T(s)$ equal to infinity. These values must therefore be those values of s which make the denominator equal to zero. Thus, from expression (2.4), the poles of $T(s)$ are $-p_1$, $-p_2$, ... and $-p_m$.

There may also be additional poles or zeros at infinity if $n \neq m$, but these are not usually required for analysis except in some root locus work (Chapters 5 and 6).

Example 2.1. Two systems have transfer functions given by:

$$\text{i.} \quad T_1(s) = \frac{4s^2 + 8s}{s^2 + 11s + 28} \, . \qquad ...(2.6)$$

$$\text{ii.} \quad T_2(s) = \frac{s^2 + 2500s + 10^6}{10s^4 + 1000s^3 + 125\,000s^2} \, . \qquad ...(2.7)$$

In each case, rearrange the expression to obtain the *PZ* form and hence, identify the poles and zeros of the transfer function.

Solution. (i) The first step is to find the common factor K which reduces the coefficients of the highest powers of s to unity:

$$T_1(s) = \frac{4\,(s^2 + 2s)}{s^2 + 11s + 28} \, . \qquad ...(2.8)$$

The numerator and denominator are then factorized:

$$T_1(s) = \frac{4s\,(s+2)}{(s+4)\,(s+7)}\,. \qquad \ldots(2.9)$$

Note, the factorization of quadratic expressions can be achieved by the use of the formula:

if,

$$as^2 + bs + c = 0\,,$$

then,

$$s = \frac{-b \pm \sqrt{b^2 - 4ac}}{2a}\,. \qquad \ldots(2.10)$$

But in the procedures used here, a is always reduced to unity and the formula may be rewritten:

$$s = \frac{-b}{2} \pm \sqrt{\left(\frac{b}{2}\right)^2 - c}\,. \qquad \ldots(2.11)$$

This form is simpler and has the advantage that the roots are more easily obtained with the aid of a simple calculator with a single memory.

In this example, put

$$s^2 + 11s + 28 = 0,$$

applying the form in (2.11),

$$s = \frac{-11}{2} \pm \sqrt{\frac{121}{4} - 28}, = -4 \text{ or } -7\,.$$

These are the values of s which will make $s^2 + 11s + 28 = 0$. They are also the values of s which make $(s+4)(s+7) = 0$.

$$\therefore \quad s^2 + 11s + 28 = (s+4)(s+7)\,.$$

Now, returning to expression (2.9), the zeros are the values of s which make $T_1(s) = 0$. By inspection, these are $s = 0$ and $s = -2$. The poles of $T_1(s)$ are the values of s which make $T_1(s) = \infty$ or the values of s which make the denominator of $T_1(s) = 0$. By inspection, these are $s = -4$ and $s = -7$.

Solution. (ii) Using the same steps,

$$T_2(s) = \frac{0.1\,(s^2 + 2500s + 10^6)}{s^2\,(s^2 + 100s + 12\,500)}\,. \qquad \ldots(2.12)$$

To find the factors of the numerator,

$$s = -1250 \pm \sqrt{1250^2 - 10^6} = -500 \text{ or } -2000\,.$$

For the factors of the denominator,

$$s = -50 \pm \sqrt{2500 - 12\,500}\,.$$

In this case, the number under the square root is negative,

$$\therefore \quad s = -50 \pm j100\,,$$

$$\therefore \quad T_2(s) = \frac{0.1\,(s+500)\,(s+2000)}{s^2(s+50-j100)\,(s+50+j100)}\,. \qquad \ldots(2.13)$$

Note that the order of the two factors corresponds with the preceding expression. $s = -50$ plus or minus j100, so the factors are $(s + 50 - j100)$ and $(s + 50 + j100)$. This is important only because it will be convenient to identify the first and second factors at a later stage.

The finite zeros of $T_2(s)$ are $s = -500$ and $s = -2000$. Two of the poles of $T_2(s)$ are $s = 0$ and the other two are $s = -50 + j100$ and $s = -50 - j100$. There will be two further zeros when $s = \infty$ because as s becomes very large, the numerical parts of the factors become negligible. Thus as s tends to infinity, $T_2(s)$ tends to $\frac{0.1s^2}{s^4}$ or $\frac{0.1}{s^2}$. This result becomes zero as $s = \infty$. Since the term is s^2, there are two additional zeros at infinity. Once again, note that this only becomes important in some root locus work in later chapters.

2.5 DETERMINATION OF SYSTEM TRANSFER FUNCTIONS

Various techniques can be used to determine transfer functions, but in general, three stages are required. The first step is to 'describe' the system in terms of *s*. System equations are then formed using familiar techniques. Finally, the result is rearranged to provide the transfer function in the required form. These techniques are best illustrated by a range of examples for different classes of system.

2.6 TRANSFER FUNCTIONS FOR SIMPLE ELECTRICAL CIRCUITS

Example 2.2. Determine the transfer function in *PZ* form for each of the three networks shown in Fig. 2.1. The input and output in each case are the voltages shown.

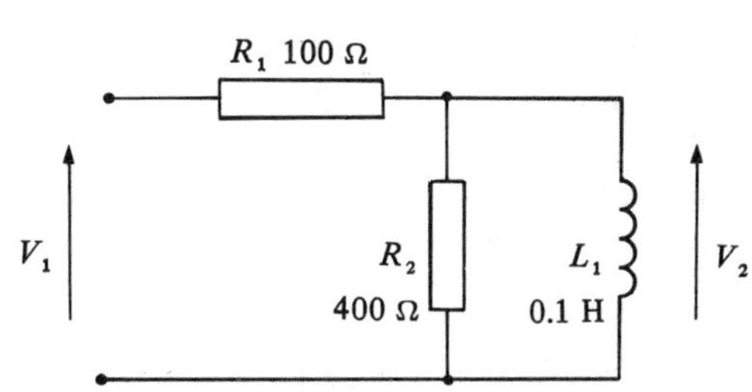

(a)

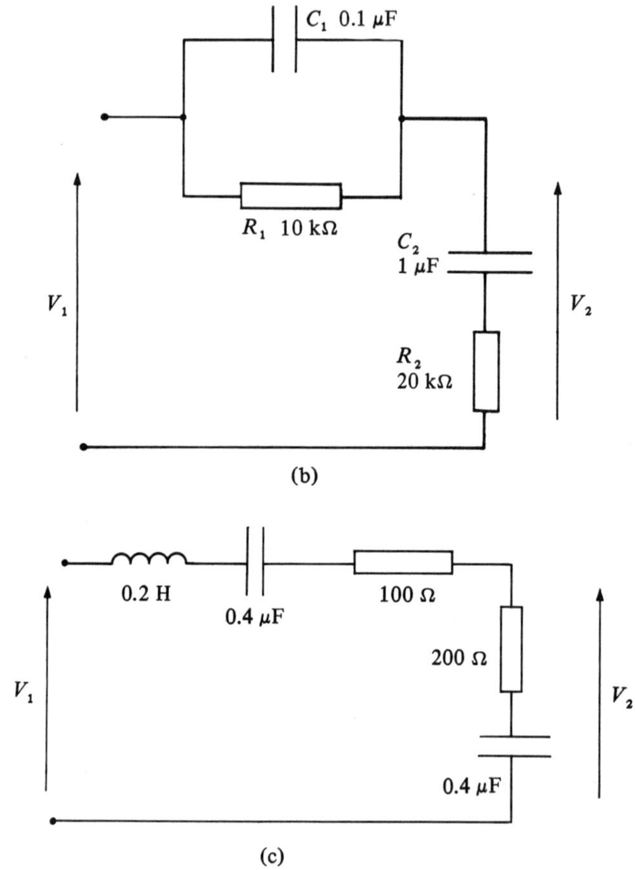

Fig. 2.1 Electrical networks for Example 2.2.

Solution. In each case, the *s* plane circuit will be drawn with numerical values inserted. The resistor values are unchanged, the inductive impedance is *sL* ohms and the capacitive impedance is $1/sC$ ohms.

a) The *s* plane circuit for the first network is shown in Fig. 2.2. This may be solved by using series-parallel relationships.

$$T(s) = \frac{Z_2}{Z_1 + Z_2} = \frac{\dfrac{40s}{400 + 0.1s}}{100 + \dfrac{40s}{400 + 0.1s}} . \qquad \ldots (2.14)$$

The next step is always to reduce the expression to a simple fraction. In this case, both numerator and denominator are multiplied by $400 + 0.1s$.

Hence,

$$T(s) = \frac{40s}{40\,000 + 10s + 40s} . \qquad \ldots (2.15)$$

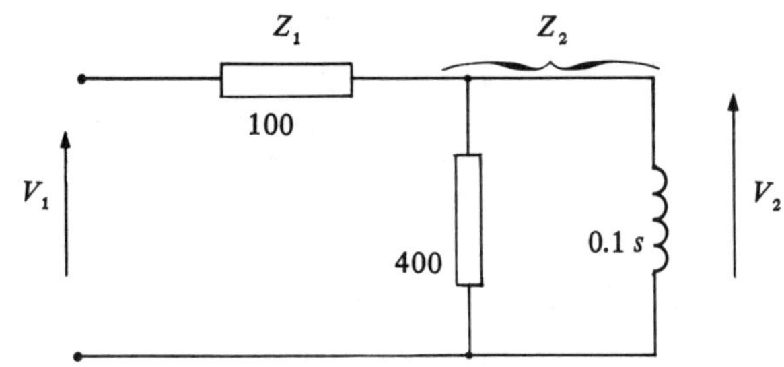

Fig. 2.2 The *s* domain circuit for Example 2.2a.

Collecting terms and rearranging,

$$T(s) = \frac{40s}{50s + 40\,000} \,. \qquad \ldots(2.16)$$

Reduce to the standard form,

$$T(s) = \frac{0.8s}{s + 800} \,. \qquad \ldots(2.17)$$

Note that the transfer function for circuit (a) has a zero at $s = 0$ and a pole at $s = -800$.

b) The *s* plane circuit for network (b) is shown in Fig. 2.3. Where a network includes several series and parallel combinations, as in this case, it is often more convenient to find the combined impedances first. In this case, the parallel combination Z_p and the series combination Z_s are found as follows:

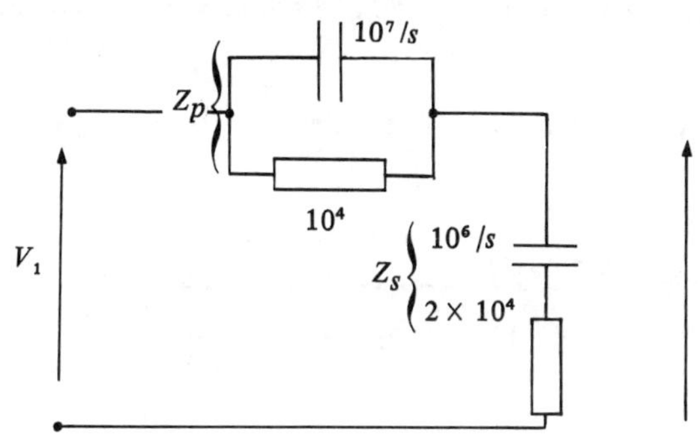

Fig. 2.3 The *s* domain circuit for Example 2.2b.

$$Z_p = \frac{10^{11}/s}{10^4 + 10^7/s} = \frac{10^{11}}{10^4 s + 10^7} = \frac{10^7}{s + 10^3} \qquad \ldots(2.18)$$

and,

$$Z_s = \frac{10^6}{s} + 2 \times 10^4 = \frac{2 \times 10^4\,(s + 50)}{s} \qquad \ldots(2.19)$$

Thus,

$$T(s) = \frac{V_2}{V_1} = \frac{\dfrac{2 \times 10^4 (s + 50)}{s}}{\dfrac{10^7}{s + 10^3} + \dfrac{2 \times 10^4 (s + 50)}{s}} \qquad \ldots (2.20)$$

Multiplying numerator and denominator by $(s + 10^3)$ and by s:

$$T(s) = \frac{2 \times 10^4 (s + 50)(s + 10^3)}{10^7 s + 2 \times 10^4 (s + 50)(s + 10^3)} \qquad \ldots (2.21)$$

$$= \frac{2 \times 10^4 (s + 50)(s + 10^3)}{10^7 s + 2 \times 10^4 (s^2 + 1050s + 5 \times 10^4)} \qquad \ldots (2.22)$$

Dividing through by 2×10^4 and collecting terms,

$$T(s) = \frac{(s + 50)(s + 10^3)}{s^2 + 1550s + 5 \times 10^4} \qquad \ldots (2.23)$$

Factorizing,

$$T(s) = \frac{(s + 50)(s + 10^3)}{(s + 33)(s + 1517)} \qquad \ldots (2.24)$$

This transfer function has two zeros at $s = -50$ and $s = -1000$ and two poles at $s = -33$ and at $s = -1517$.

c) The s plane circuit is shown in Fig. 2.4. Working directly from the diagram,

$$T(s) = \frac{V_2(s)}{V_1} = \frac{200 + \dfrac{2.5 \times 10^6}{s}}{0.2s + 300 + \dfrac{5 \times 10^6}{s}} \,. \qquad \ldots (2.25)$$

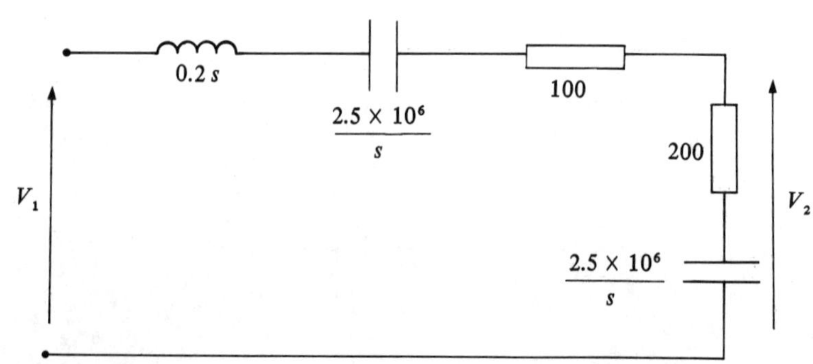

Fig. 2.4 The s domain circuit for Example 2.2c.

Multiplying through by s,

$$T(s) = \frac{200s + 2.5 \times 10^6}{0.2s^2 + 300s + 5 \times 10^6}\,. \qquad \ldots (2.26)$$

Reducing the coefficients of the highest powers of s to unity,

$$T(s) = \frac{1000\,(s + 1.25 \times 10^4)}{s^2 + 1500s + 2.5 \times 10^7}\,. \qquad \ldots (2.27)$$

Factorizing the denominator,

$$T(s) = \frac{1000\,(s + 1.25 \times 10^4)}{(s + 750 - \mathrm{j}4943)\,(s + 750 + \mathrm{j}4943)}\,. \qquad \ldots (2.28)$$

This transfer function has a real zero at $s = -1.25 \times 10^4$ but the two poles are a complex conjugate pair at $s = -750 + \mathrm{j}4943$ and $s = -750 - \mathrm{j}4943$.

2.7 GENERAL RULES FOR OBTAINING SIMPLE CIRCUIT TRANSFER FUNCTIONS

In the last example, all three networks have numerical values. This makes the evaluation of transfer functions easier than for general cases using symbols for unspecified components. Before proceeding to such cases, it may be useful to summarize the steps used in the last example, to obtain the circuit transfer function in the pole zero form.

i. Express all impedances or admittances as functions of s and select the required voltage and current variables.
ii. Write circuit equations to obtain an expression for the transfer function.
iii. Eliminate fractions in both numerator and denominator by multiplying both by the necessary factors. More than one step may be required for this reduction.
iv. Multiply out brackets if necessary and collect terms in powers of s.
v. Reduce the coefficients of the highest powers of s in both numerator and denominator to unity.
vi. Factorize numerator and denominator to leave the transfer function in the pole zero form.

2.8 TRANSFER FUNCTIONS FOR GENERAL CASES

In the next example, we see how general cases can be manipulated and how a specific circuit can be modified so that the pole and zero frequencies can be changed without affecting their relative values.

Example 2.3. Determine the transfer functions for the two circuits shown in Fig. 2.5 and in each case, find new values for the components so that the lowest pole frequency is –500 while the relative values of the other poles and zeros is unchanged.

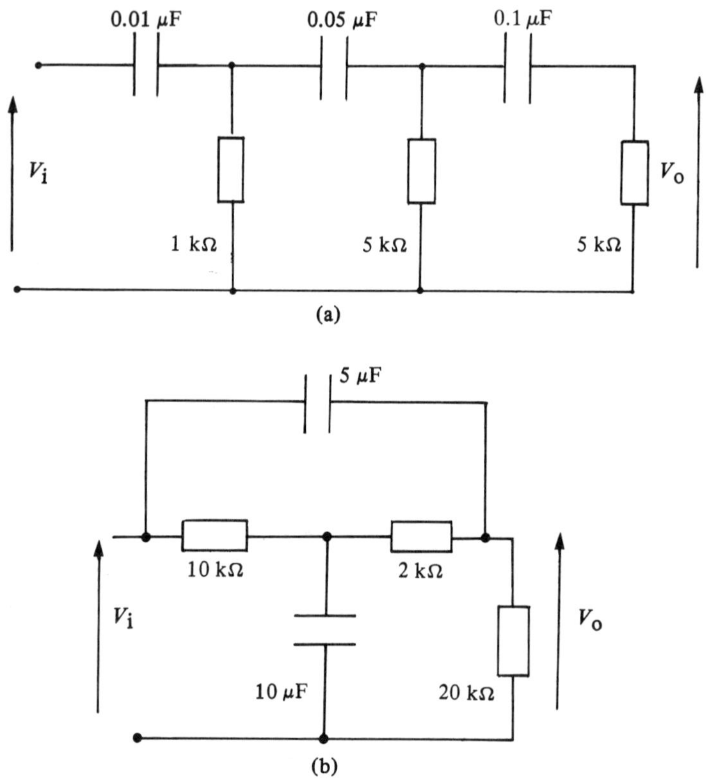

Fig. 2.5 Electrical networks for Example 2.3; analysis by mesh and nodal analysis.

Solution. The first circuit is in a form that is conveniently solved using mesh analysis with three circulating unknown currents. The same method could be adopted for the second circuit, but less work is involved if nodal analysis is employed as there are then only two unknown voltages compared with the three unknown currents required for mesh analysis. Returning to the first circuit, the problem can be made more general by taking single values of C and R and then referring the others to these. The two values will be 1 kΩ and 0.1 μF. The circuit is then redrawn in the s plane as shown in Fig. 2.6.

Writing the equations for the three meshes by inspection:

$$V_i = I_1(10/sC + R) - I_2R. \qquad \ldots(2.29)$$

$$0 = -I_1R + I_2(2/sC + 6R) - I_3 5R \qquad \ldots(2.30)$$

$$0 = -5RI_2 + I_3(1/sC + 10R) \qquad \ldots(2.31)$$

Multiply all the equations by sC to eliminate the fractions,

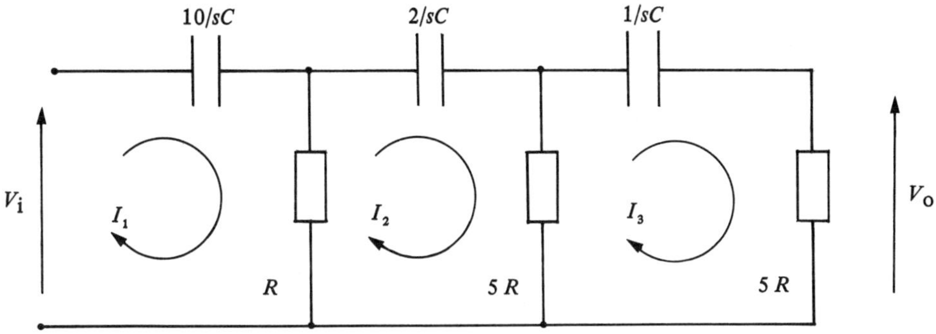

Fig. 2.6 The s domain circuit for Example 2.3a.

$$sCV_i = I_1(10 + sCR) - I_2 sCR \qquad \ldots(2.32)$$

$$0 = -I_1 sCR + I_2(2 + 6sCR) - I_3 5sCR \qquad \ldots(2.33)$$

$$0 = -I_2 5sCR + I_3(1 + 10sCR) \qquad \ldots(2.34)$$

Solving by determinants,

$$I_3 = \frac{sCV_i \times 5s^2C^2R^2}{(10 + sCR)(2 + 6sCR)(1 + 10sCR) - 25s^2C^2R^2 - s^2C^2R^2(1 + 10sCR)} \qquad \ldots(2.35)$$

$$= \frac{5s^3C^3R^3V_i}{25s^3C^3R^3 + 375s^2C^2R^2 + 262sCR + 20} \qquad \ldots(2.36)$$

Dividing through by $25C^3R^3$ and noting that $V_o = 5000I_3$,

$$T(s) = \frac{V_o}{V_i} = \frac{1000s^3}{s^3 + \dfrac{15s^2}{CR} + \dfrac{10.48s}{C^2R^2} + \dfrac{0.8}{C^3R^3}}. \qquad \ldots(2.37)$$

As the poles of the transfer function are to be determined, the cubic in the denominator must be factorized. (One method of factorizing higher-order polynomials is demonstrated in Chapter 6.) The result then becomes:

$$T(s) = \frac{1000s^3}{(s + \dfrac{0.64}{CR})(s + \dfrac{0.09}{CR})(s + \dfrac{14.27}{CR})}. \qquad \ldots(2.38)$$

Thus for the circuit given, as CR is 10^{-4}, the poles are at -900, -6400 and -14.27×10^4; the three zeros are all of value 0. If the lowest pole is to be -500, then, $\dfrac{0.09}{CR} = 500$, or $CR = 1.8 \times 10^{-4}$. This condition would be satisfied by leaving the capacitors unchanged and using resistors of 1.8 kΩ and 9 kΩ respectively.

For the second circuit, we work with general values, but as nodal analysis is used, the components are more conveniently expressed as admittances. The required circuit is shown in Fig. 2.7 with the unknown node voltages indicated as V_1, V_2 and V_3.

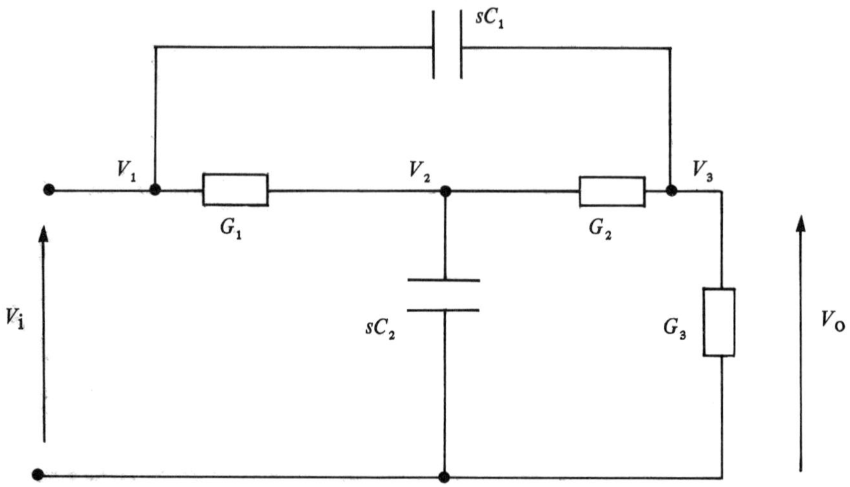

Fig. 2.7 The *s* domain circuit for Example 2.3b.

Writing two nodal equations directly from the circuit;

$$(V_1 - V_2)G_1 = V_2(G_2 + sC_2) - V_3G_2\,, \qquad \ldots(2.39)$$

$$(V_1 - V_3)sC_1 = -V_2G_2 + V_3(G_2 + G_3)\,; \qquad \ldots(2.40)$$

rearranging,

$$G_1V_1 = V_2(G_1 + G_2 + sC_2) - V_3G_2\,, \qquad \ldots(2.41)$$

$$sC_1V_1 = -V_2G_2 + V_3(G_2 + G_3 + sC_1)\,. \qquad \ldots(2.42)$$

Solving by determinants,

$$V_3 = \frac{sC_1V_1(G_1 + G_2 + sC_2) + G_1G_2V_1}{(G_1 + G_2 + sC_2)(G_2 + G_3 + sC_1) - G_2^2}\,. \qquad \ldots(2.43)$$

$$\therefore\ \frac{V_3}{V_1} = \frac{s^2C_1C_2 + s(C_1G_1 + C_1G_2) + G_1G_2}{s^2C_1C_2 + s(C_2G_2 + C_2G_3 + C_1G_1 + C_1G_2) + (G_1G_2 + G_1G_3 + G_2G_3)} \qquad \ldots(2.44)$$

Dividing through by C_1C_2, and writing $G = 1/R$,

$$T(s) = \frac{V_o}{V_i} = \frac{s^2 + s\left(\frac{1}{C_2R_1} + \frac{1}{C_2R_2}\right) + \frac{1}{C_1C_2R_1R_2}}{s^2 + s\left(\frac{1}{C_1R_2} + \frac{1}{C_1R_3} + \frac{1}{C_2R_1} + \frac{1}{C_2R_2}\right) + \frac{1}{C_1C_2}\left(\frac{1}{R_1R_2} + \frac{1}{R_1R_3} + \frac{1}{R_2R_3}\right)}\,. \qquad \ldots(2.45)$$

This is a reassuring answer since R and C values occur in pairs – giving rise to time constants, which, combined with the dimensions of s, ensure $T(s)$ is dimensionless.

It would serve no useful purpose to factorize the expressions in the numerator and the denominator, but if we insert the values given, the above result becomes:

$$T(s) = \frac{s^2 + 60s + 1000}{s^2 + 170s + 1600} . \qquad \ldots (2.46)$$

Factorizing,

$$T(s) = \frac{(s + 30 - j10)(s + 30 + j10)}{(s + 10)(s + 160)} . \qquad \ldots (2.47)$$

This result is interesting as, although the two poles are real and negative with s equal to -10 and -160, the zeros are a complex conjugate pair with values $-30 + j10$ and $-30 - j10$.

In order to find the required new component values, it is convenient to put $C_1 = C$ and $R_2 = R$. This results in $C_2 = 2C$, $R_1 = 5R$ and $R_3 = 10R$. If the reader substitutes these values into result (2.45) above, he can show that the poles have values $\frac{-0.1}{CR}$ and $\frac{-1.6}{CR}$ and the zeros are at $\frac{-0.3 + j0.1}{CR}$ and $\frac{-0.3 - j0.1}{CR}$. Using these results, suitable component values can be found to produce the required lower pole frequency of -500. One solution leaves the resistor values unchanged with capacitors of 0.1 μF and 0.2 μF respectively.

2.9 TRANSFER FUNCTIONS OF NETWORKS INVOLVING AMPLIFIERS

Amplifiers can be used in many ways to realize different forms of transfer function. In simple arrangements, amplifiers may be used to isolate one network from another as the amplifier input and output impedances can approximate to infinity and zero respectively. Thus, each isolated network can be treated separately since the amplifier prevents network loading by drawing negligible current ($z_{IN} = \infty$) and it provides an almost ideal voltage source to the next network ($z_{OUT} = 0$). The voltage amplification in such cases will often be unity (with or without phase inversion) but other values of gain can be used if required. Alternatively, amplifiers can be used to produce the sum or the difference between two voltages while still providing isolation and gain if required. Further advantages are obtained by the use of high-gain inverting amplifiers in the 'operational' mode. It may be important to know that all amplifiers will add poles to transfer functions (but usually no zeros with d.c. amplifiers). In some cases, these poles will be at very high frequencies and their effects can be ignored for many applications. With other types of amplifier, the effect of the very high frequency poles is 'eclipsed' for all practical purposes by a single, well-defined pole at a relatively low frequency. The next examples demonstrate the use of isolating and difference amplifiers.

Example 2.4. In the circuit shown in Fig. 2.8, the amplifiers are voltage followers. Determine the voltage transfer function $\frac{V_o}{V_i}(s)$ and express the result in the Bode form.

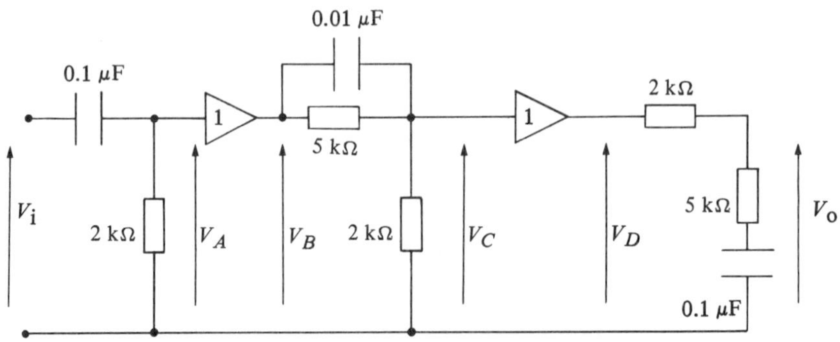

Fig. 2.8 A circuit using a voltage follower for Example 2.4.

Solution. Since the amplifiers are voltage followers, each will have a gain of one and input and output impedances approximating to infinity and zero respectively.

The transfer function required can be expressed in the following way:

$$\frac{V_o}{V_i} = \frac{V_A}{V_i} \times \frac{V_B}{V_A} \times \frac{V_C}{V_B} \times \frac{V_D}{V_C} \times \frac{V_o}{V_D} \qquad \ldots (2.48)$$

where,

$$\frac{V_B}{V_A} = \frac{V_D}{V_C} = 1$$

and the remaining component transfer functions can be calculated using the simple series parallel relationships shown in Example 2.2. The reader can confirm that these are;

$$\frac{V_A}{V_i} = \frac{s}{s + 5000},$$

$$\frac{V_C}{V_B} = \frac{s + 20\,000}{s + 70\,000}, \quad \text{and} \quad \frac{V_o}{V_D} = \frac{0.714\,(s + 2000)}{(s + 1430)}. \qquad \ldots (2.49)$$

Combining and rearranging in the Bode form results in:

$$T(s) = \frac{5.7 \times 10^{-5}\, s\,(1 + 0.5s)\,(1 + 0.05s)}{(1 + 0.7s)\,(1 + 0.2s)\,(1 + 0.014s)}, \qquad \ldots (2.50)$$

where the time constants are expressed in milliseconds.

Example 2.5. In Fig 2.9, the amplifier A_1 is a summing amplifier with a gain of one. The summing effect is indicated by the summing junction symbol. The amplifier A_2 is phase inverting and has variable gain. Determine the system poles and zeros if the magnitude of the gain of A_2 is set to (a) 0.1 or (b) 1.0.

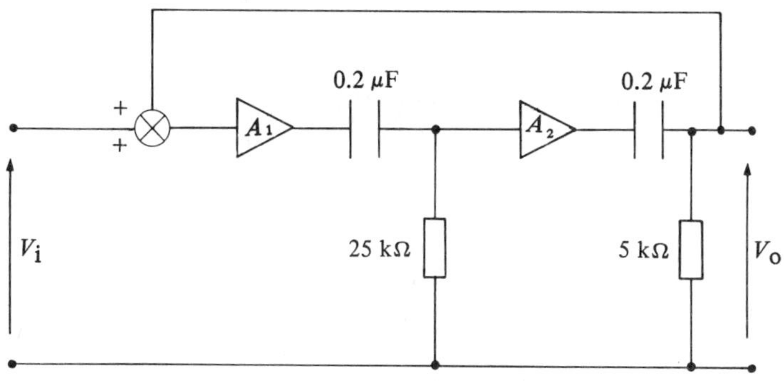

Fig. 2.9 A feedback amplifier circuit for Example 2.5.

Solution. In this example, feedback is applied. Since the two signals are *added* at the input, the familiar formula $\dfrac{A}{1-BA}$ is applicable, where A is the forward gain without feedback and, in this case, $B = 1$. In both cases, there are two simple CR potential dividers with transfer functions of $\dfrac{s}{s + 1/CR}$. Inserting values, we will find that A is given by:

$$A = \frac{A_2 s^2}{(s + 200)\,(s + 1000)} = \frac{A_2 s^2}{s^2 + 1200s + 2 \times 10^5}. \qquad \ldots (2.51)$$

Substituting in the feedback formula;

$$T(s) = \frac{\dfrac{A_2 s^2}{s^2 + 1200s + 2 \times 10^5}}{1 - \dfrac{A_2 s^2}{s^2 + 1200s + 2 \times 10^5}}. \qquad \ldots (2.52)$$

$$T(s) = \frac{A_2 s^2}{s^2 + 1200s + 2 \times 10^5 - A_2 s^2}. \qquad \ldots (2.53)$$

$$T(s) = \frac{A_2 s^2}{s^2(1 - A_2) + 1200s + 2 \times 10^5}. \qquad \ldots (2.54)$$

Note that in this system, the pole positions are determined, not only by the values of the passive components, but also by the gain of an amplifier.

Substituting for the first value of A_2;

$$T(s) = \frac{-0.1s^2}{1.1s^2 + 1200s + 2 \times 10^5}, \qquad \ldots (2.55)$$

$$= \frac{-0.09s^2}{(s + 206)\,(s + 884)}. \qquad \ldots (2.56)$$

Substituting for the second value of A_2 in the same way, results in a complex conjugate pair of poles at $s = -300 + j100$ and $-300 - j100$.

This result is also of interest as previous examples involving only *RC* networks (or only *RL* networks) have resulted in transfer functions with only *real negative* poles. This example shows that complex conjugate pairs of poles, which can occur with *RLC* networks, can also be obtained using *RC* networks in a feedback system. The usefulness of complex conjugate poles will be seen later, but they basically imply highly selective networks and therefore this example has two important implications.

i. Highly selective, active *RC* networks can be realized without the need for inductors – making them suitable for miniaturization in integrated circuit form.
ii. The use of isolating amplifiers suggests a method of synthesis, where, given a transfer function, a network can be realized in a simple manner.

2.10 TRANSFER FUNCTIONS OF OPERATIONAL AMPLIFIER ARRANGEMENTS

Operational amplifiers are inverting amplifiers having a very high gain and input and output impedances which approximate to infinity and zero respectively. Most commercially available 'op-amps' have, in addition, differential inputs allowing for non-inverting amplification or the amplification of difference signals if required. They are used in many different arrangements with passive components, but two of the most important within the context of this chapter are shown in Fig. 2.10.

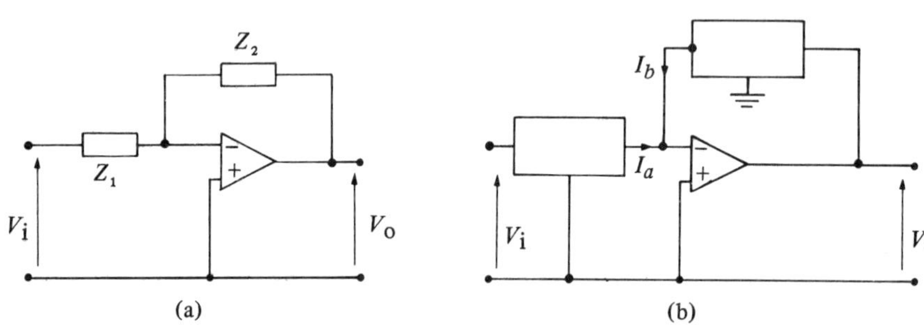

Fig. 2.10 Some common operational amplifier arrangements.

Circuit (a) is probably familiar to most readers and the gain or transfer function is given by:

$$\frac{V_o}{V_i} = \frac{-Z_2}{Z_1} \qquad \ldots(2.57)$$

This form is convenient if Z_1 and Z_2 are single components or series combinations of components. If either branch includes parallel components, one of the following forms will be more useful.

$$\frac{V_o}{V_i} = \frac{-Y_1}{Y_2} = -Z_2Y_1 = \frac{-1}{Z_1Y_2} . \qquad \ldots(2.58)$$

Examples demonstrating the alternative use of these forms will be found below.

Referring to the circuit in Fig. 2.10b, the branches Z_1 and Z_2 are replaced by three terminal networks. The familiar virtual earth concept still applies at the inverting input to the amplifier. This means that there will be zero potential between the inverting terminal and earth, and no current flows into the high impedance input terminal. The two currents I_a and I_b are flowing into an apparent open-circuit; there is no other current path, so $I_a = -I_b$. But,

$$I_a = -V_1 y_{21a} ,$$

where y_{21a} is the forward transfer admittance parameter of the input network.

$$I_b = -V_o y_{12b} ,$$

where y_{21b} is the reverse transfer admittance parameter of the feedback network. But, for any passive network, $y_{12} = y_{21}$.

$$\therefore \quad -V_i y_{21a} = V_o y_{21b} ,$$

or

$$\frac{V_o}{V_i} = \frac{-y_{21a}}{y_{21b}} ; \qquad \ldots(2.59)$$

which should be compared with one of the simple forms in expression (2.58). The form $\frac{-Y_1}{Y_2}$ is, in fact, identical to result (2.59), as y_{21} for a two-terminal network is simply the admittance of that network.

Finally, the transfer parameter for any three-terminal network can be determined from:

$$y_{21} = \left.\frac{I_2}{V_1}\right|_{V_2 = 0} . \qquad \ldots(2.60)$$

Thus for any particular network, the output is short-circuited to make $V_2 = 0$, and then, with a voltage V_1 applied, I_2, the current flowing from the short-circuit into the network, can be determined.

Example 2.6. Determine the voltage transfer functions for the two operational amplifier circuits shown in Fig. 2.11.

Solution. For circuit (a), $Z_1 = R, Z_2 = \frac{1}{sC}$, or $Y_2 = sC$, hence,

$$T(s) = \frac{V_o}{V_i} = \frac{-1}{Z_1Y_2} = \frac{-1}{sCR} . \qquad \ldots(2.61)$$

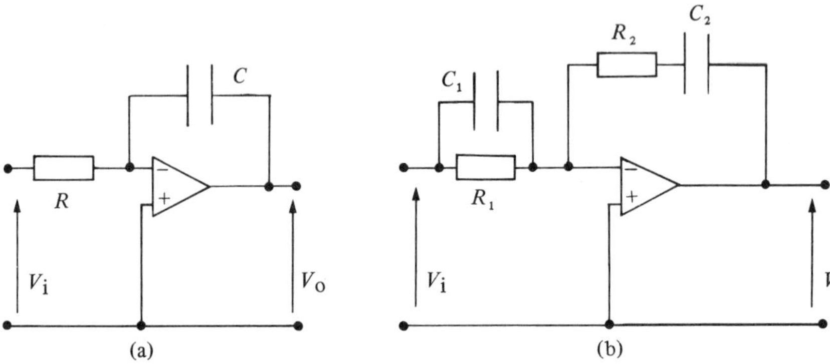

Fig. 2.11 Operational amplifier circuits for Example 2.6.

Thus, this arrangement produces no zeros and a pole at $s = 0$ which results in V_o being the integral of V_i.

For circuit (b), C_1 and R_1 are in parallel making $Y_1 = 1/R_1 + sC_1$. C_2 and R_2 are in series, so we use $Z_2 = R_2 + 1/sC_2$. Hence,

$$T(s) = -Z_2 Y_1 = -(R_2 + 1/sC_2)(1/R_1 + sC_1). \qquad \ldots(2.62)$$

This result can be rearranged to:

$$T(s) = \frac{-C_1R_2(s + 1/C_1R_1)(s + 1/C_2R_2)}{s}. \qquad \ldots(2.63)$$

This transfer function has a pole at $s = 0$ and two real zeros whose values can be selected independently.

Example 2.7. Calculate the transfer parameter y_{21} for each of the two three-terminal networks shown in Fig. 2.12a and hence determine the poles and zeros of the voltage transfer function for the complete arrangement.

Solution. Fig. 2.12b shows the input network redrawn for the calculation of y_{21a}. To obtain the more general form, the admittance of the 10 kΩ resistors and the 0.02 μF capacitors are shown as G and sC respectively. In the same way, the components of the feedback network shown in Fig. 2.12c are shown as $2G$ (for the 5 kΩ resistors) and sC.

From Fig. 2.12b,

$$-I_2 = V_1 \times \frac{G(G + sC)}{(2G + sC)} \times \frac{G}{(G + sC)}. \qquad \ldots(2.64)$$

$$\therefore \quad y_{21a} = \frac{I_2}{V_1} = \frac{-G^2}{2G + sC}. \qquad \ldots(2.65)$$

Substituting for $G = 1/R$ and rearranging,

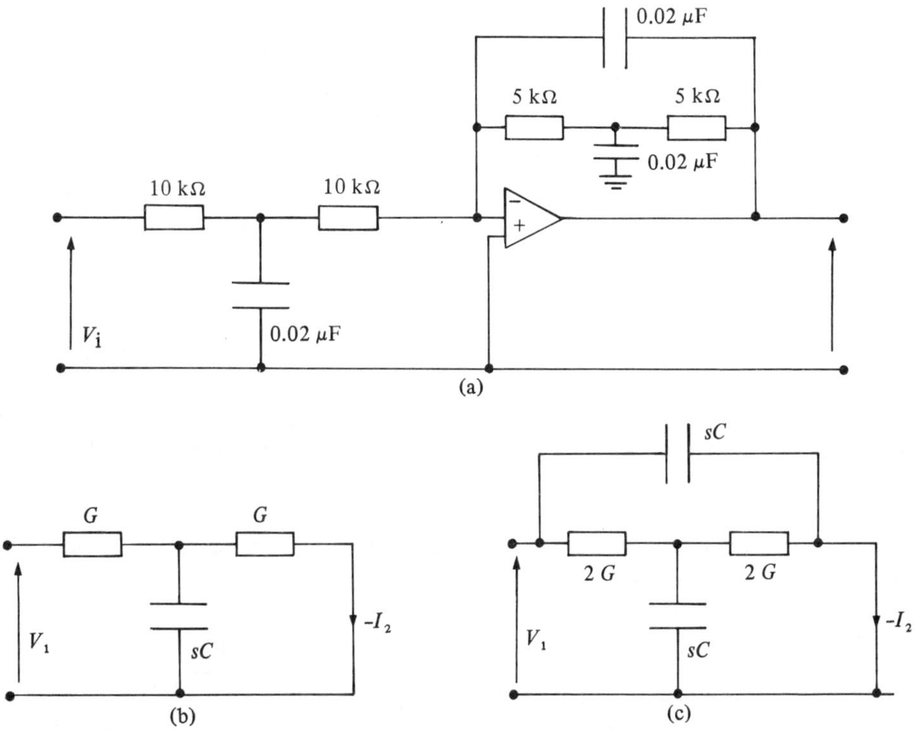

Fig. 2.12 Operational amplifier circuit with three terminal networks. a) The complete circuit. b) The *s* domain input network. c) The *s* domain feedback network.

$$y_{21a} = \frac{-1/R^2C}{(s + 2/CR)} \; . \qquad \ldots(2.66)$$

Referring to Fig. 2.12c,

$$-I_2 = V_1 \frac{2G(2G + sC)}{(4G + sC)} \times \frac{2G}{(2G + sC)} + sC \; ; \qquad \ldots(2.67)$$

$$y_{21b} = \frac{I_2}{V_1} = -\frac{4G^2 + sC(4G + sC)}{4G + sC} \; . \qquad \ldots(2.68)$$

Substituting for $G = 1/R$ and rearranging,

$$y_{21b} = \frac{-C(s^2 + 4s/CR + 4/C^2R^2)}{(s + 4/CR)} \; . \qquad \ldots(2.69)$$

Substituting the values for C and R makes $1/CR = 5000$, hence,

$$y_{21a} = \frac{-0.5}{s + 10^4} \; , \qquad \ldots(2.70)$$

and

$$y_{21b} = \frac{-2 \times 10^{-8}\,(s^2 + 2 \times 10^4 s + 10^8)}{(s + 2 \times 10^4)} \qquad \ldots(2.71)$$

The numerator of (2.71) is a perfect square and may be written $(s + 10^4)^2$. Now applying result (2.59) and substituting (2.70) and (2.71) to obtain the system transfer function;

$$T(s) = -\frac{y_{21a}}{y_{21b}} = \frac{-0.25 \times 10^8\,(s + 2 \times 10^4)}{(s + 10^4)^3} \qquad \ldots(2.72)$$

Thus this system has a single zero at $s = -2 \times 10^4$ and three coincident poles at $s = -10^4$. In practice, with component tolerances, the three poles would have slightly different values but, as we see in later chapters, the effects of such discrepancies on the transient and steady-state properties would be negligible. As a final example in this section, we shall consider an 'op-amp' arrangement using one two-terminal network and one three-terminal network.

Example 2.8. Determine the voltage transfer function for the scaled circuit shown in Fig. 2.13.

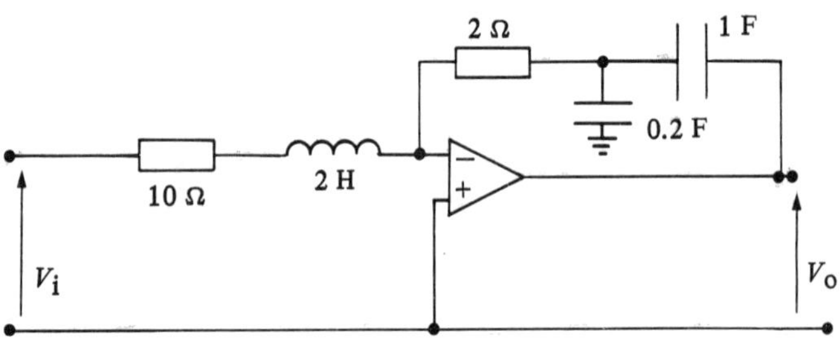

Fig. 2.13 The operational amplifier circuit for Example 2.8.

Solution. In this circuit, $-y_{21a}$ is the admittance of the input network, but as this has a series arrangement, it is more convenient to express it as:

$$-y_{21a} = -\frac{1}{Z_a} \qquad \ldots(2.73)$$

The feedback network is in the three-terminal form encountered in the last example. The required transfer function is thus given by:

$$T(s) = \frac{-1}{Z_a y_{21b}} \qquad \ldots(2.74)$$

It is left to the reader to show that the final result is:

$$T(s) = \frac{1.2\,(s + 0.416)}{s\,(s + 5)} \qquad \ldots(2.75)$$

2.11 ACTIVE FILTER CIRCUITS

Alternative arrangements are commonly encountered in active filter circuits. The writing of nodal equations is often the best approach to analysis and variations on this are illustrated in the next example.

Example 2.9. Three active filter circuits are shown in Fig. 2.14. In each case, write a set of equations defining the circuit action and hence solve for the voltage transfer function.

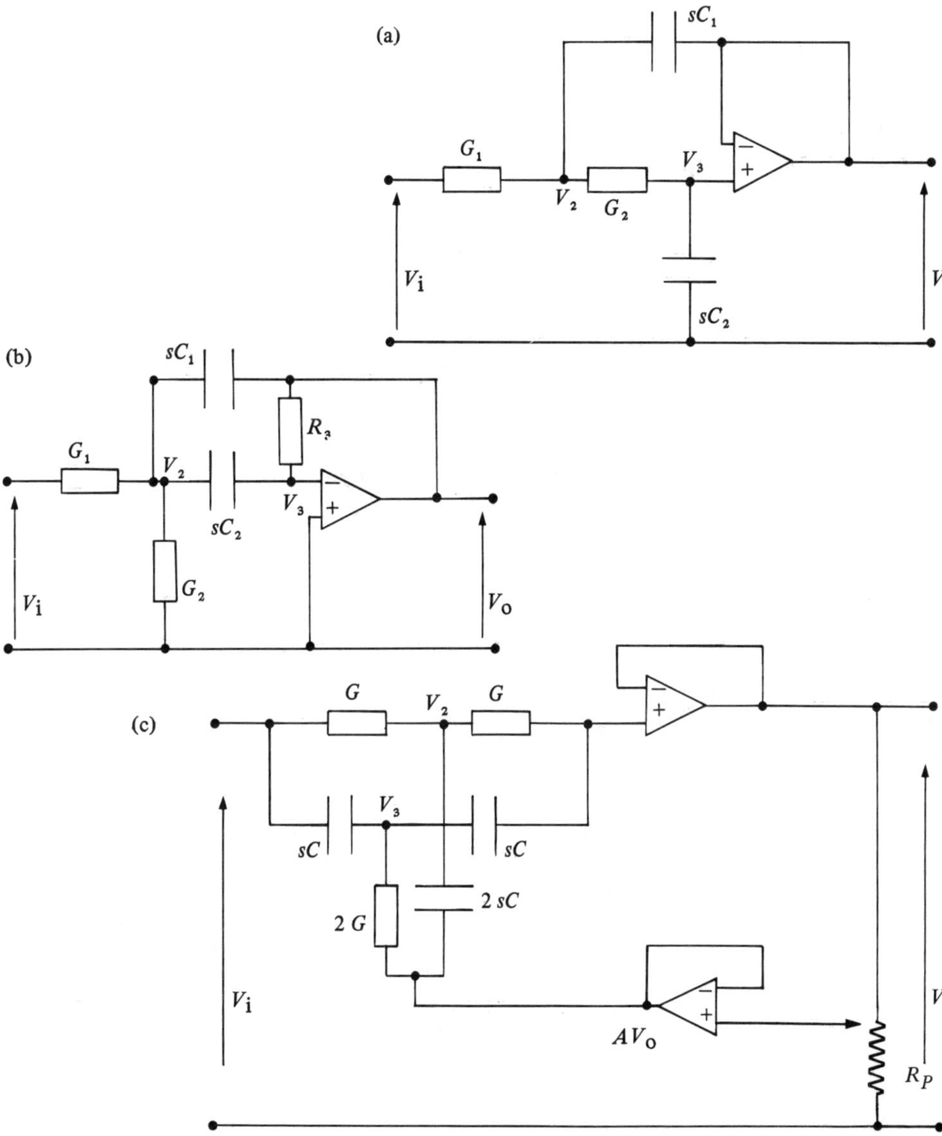

Fig. 2.14 Some active filter circuits using operational amplifiers. a) Low pass. b) Dual feedback band-pass. c) Band stop with variable selectivity.

Solution. a) In this circuit, the 'op-amp' is used as a voltage follower making $V_o = V_3$. Thus, there is one applied voltage V_1, and two unknown voltages V_2 and V_3. Two nodal equations can be written and solved to give V_3 or V_o in terms of V_1. As nodal analysis is to be used, the components are more conveniently expressed in the admittance form as shown. The required equations are:

$$(V_1 - V_2)\,G_1 = V_2\,(G_2 + sC_1) - V_3G_2 - V_osC_1\,; \qquad \ldots(2.76)$$

$$0 = -V_2G_2 + V_3\,(G_2 + sC_2)\,. \qquad \ldots(2.77)$$

Substituting $V_3 = V_o$, solving and rearranging results:

$$T(s) = \frac{V_o}{V_1} = \frac{1/C_1C_2R_1R_2}{s^2 + s\left(\dfrac{1}{C_1R_1} + \dfrac{1}{C_1R_2}\right) + \dfrac{1}{C_1C_2R_1R_2}}\,. \qquad \ldots(2.78)$$

Practical values for the capacitors and resistors result in a pair of complex conjugate poles; the importance of this is discussed in Chapter 4.

b) This arrangement is known as a dual feedback circuit and it may be used as an active bandpass filter or as an active tuned circuit. The amplifier here is used with negative feedback through R_3 so that a virtual earth appears at the inverting input. As a result, the current flowing in C_2 is given by sC_2V_2. The input admittance to the amplifier is negligible, so all this current flows through R_3 to the output. The output voltage is thus $sC_2V_2 \times -R_3$, or,

$$V_2 = \frac{V_o}{sC_2R_3}\,. \qquad \ldots(2.79)$$

The other equation required is written for node V_2.

$$(V_1 - V_2)G_1 = V_2\,(G_2 + sC_1 + sC_2) - V_osC_1 \qquad \ldots(2.80)$$

Substituting for V_2, solving and rearranging, results in:

$$T(s) = \frac{V_o}{V_i} = \frac{-s/C_1R_1}{s^2 + s\left(\dfrac{C_1 + C_2}{C_1C_2}\right)\dfrac{1}{R_3} + \left(\dfrac{1/R_1 + 1/R_2}{C_1C_2R_3}\right)}\,. \qquad \ldots(2.81)$$

Again a pair of complex conjugate poles are also provided with a zero at the origin. This is the same arrangement of poles and zeros as that provided by a series *LCR* circuit, the effects of which are shown in the next two chapters.

c) In this circuit, both amplifiers are used as voltage followers (infinite input impedance and non-inverting unity gain). One, however, is used in a feedback path with a potentiometer at the input controlling the gain A to values between 0 and 1. There are three unknown voltages; V_2, V_3 and V_o, and the applied voltage is V_i. Two nodal equations can be written for the V_2 and V_3 nodes as in the previous circuits. For the third equation, note that the non-inverting input to the forward amplifier is equal to V_o and that no current flows into this input.

The writing of the three equations and the subsequent solution by determinants is left for the reader. The special case illustrated with the ratios, C, $2C$ and R, $R/2$ results in the following transfer function.

$$T(s) = \frac{s^2 + 1/C^2R^2}{s^2 + \dfrac{4s(1-A)}{CR} + \dfrac{1}{C^2R^2}} \quad . \qquad \ldots (2.82)$$

Once again, we find a pair of poles, but in this case their values are determined by the gain setting of the potentiometer. In addition there is a pair of zeros, and their values are $\pm j/CR$. This combination will provide a bandstop filter with a selectivity determined by the setting of A.

2.12 TRANSFER FUNCTIONS FOR CONTROL SYSTEMS

Transfer function analysis is widely used in the study of closed-loop control systems. The functions of system components are studied separately and then these are combined to give the system transfer function. This can be described in terms of its poles and zeros and can be used to predict the transient and steady-state responses of the complete system. It is therefore appropriate at this stage to consider the transfer functions of some non-electrical and electro-mechanical components commonly encountered in control systems.

2.13 TRANSFER FUNCTION FOR AN INERTIA LOAD WITH VISCOUS DAMPING

In many closed-loop systems, the function of the system is either to rotate a mass or to move it in a straight line. In either case, the acceleration of the load will be opposed by friction, which as a reasonable approximation, may be assumed to be viscous (proportional to the load velocity). For the rotation case, the transfer function is determined by equating torques (Nm). A torque T applied to a load having inertia J kilogram metre squared results in an angular acceleration of $J\dfrac{d^2\theta}{dt^2}$. But the angular velocity, $\dfrac{d\theta}{dt}$ and the viscous friction F_v newton metres per radian per second produce torques opposing the acceleration. Now expressing the differentials in the Laplace form, without initial conditions:

$$T - sF_v\theta = s^2J\theta \ . \qquad \ldots (2.83)$$

Rearrange to obtain the output angle in terms of the input torque;

$$T(s) = \frac{\theta}{T} = \frac{1}{s(sJ + F_v)} = \frac{1/J}{s(s + F_v/J)} \text{ radians per newton metre} \qquad \ldots (2.84)$$

This transfer function has two poles; one at $s = 0$ and the other (a 'mechanical' pole) at $s = -F_v/J$. Notice also that the pole at $s = 0$ results from a factor $1/s$ which is equivalent to integration. This is not surprising as if a constant torque is applied, the output will continue to rotate, i.e. θ increases continually, tending to infinity.

A similar analysis for linear motion equates applied force to inertia acceleration and damping force. The resulting transfer function for position x is given by:

$$\frac{x}{F} = \frac{1/M}{s\,(s + F_v/M)} \,, \qquad \ldots (2.85)$$

which is identical in form to the previous result (2.84).

2.14 ELECTRO-MECHANICAL SYSTEMS

In position control systems, the torque is usually provided by electric motors in various configurations. When d.c. motors are used, the electrical input will either be the armature current I_a, or the field current I_f, with the other maintained at a constant value. These currents in turn will be supplied from an external voltage and be limited by the inductance and resistance of the circuit. In the case of armature current control, an additional factor is the back e.m.f. which is proportional to the velocity of the ouput.

Example 2.10. A permanent magnet d.c. motor having an output torque K_t newton metre per ampere is used to rotate a load of inertia J kilogram metres squared. The back e.m.f., proportional to the output velocity, is given by K_f volts per radian per second. The armature circuit has resistance R and inductance L. If the viscous friction is F_v newton metres per radian per second, determine the transfer function for the angular position of the output in terms of the applied voltage. It may be assumed that J and F_v include the inertia and friction due to the armature and bearings.

Solution. Two equations can be extracted from the information supplied, a voltage equation and a torque equation.

$$V - K_f s\theta = i\,(R + sL) \,. \qquad \ldots (2.86)$$

The back e.m.f. here is proportional to the output velocity $s\theta$ and is negative as it opposes the increase of current in the armature circuit. The torque equation is similar to that developed above (2.84), but in this case, the applied torque is expressed in terms of the armature current i.

$$K_t i = (s^2 J + sF_v)\,\theta \,. \qquad \ldots (2.87)$$

Substituting for i in equation (2.86) and rearranging,

$$V = \frac{s\,(sJ + F_v)\,(R + sL)\theta}{K_t} + K_f s\theta \,. \qquad \ldots (2.88)$$

Using a common denominator K_t, multiplying out and collecting terms results in;

$$T(s) = \frac{\theta}{V} = \frac{K_t/JL}{s\,(s^2 + s\,(R/L + F_v/J) + \dfrac{K_f K_t}{JL})} \,. \qquad \ldots (2.89)$$

Once again the integration appears, but in this case there are two additional poles whose values depend upon both mechanical and electrical constants.

Example 2.11. A closed-loop position control system employs a field controlled d.c. motor driven from an amplifier. The amplifier input is proportional to the difference between the output angle θ_o and a reference angle θ_i. Assuming an inertia load with viscous friction and that the effect of field circuit inductance can be neglected, develop a transfer function for the system.

Solution. We can start with the torque equation obtained from expression (2.83);

$$T = s\,(sJ + F_v)\,\theta_o \,. \qquad \ldots(2.90)$$

and $T = \dfrac{AK_m V}{R}$, where A and V are the amplifier input voltage and gain respectively, R is the field circuit resistance and K_m is the motor constant. But,

$$V = K_p\,(\theta_i - \theta_o) \,. \qquad \ldots(2.91)$$

Combining the various constants and equating, we obtain:

$$K(\theta_i - \theta_o) = s\,(sJ + F_v).\theta_o \,. \qquad \ldots(2.92)$$

Rearranging,

$$T(s) = \frac{\theta_o}{\theta_i} = \frac{K/J}{s^2 + sF_v/J + K/J} \,. \qquad \ldots(2.93)$$

This system has two poles whose values are determined by K which involves the amplifier gain. In practice, there would be additional poles resulting from the field circuit inductance and from the amplifier itself.

In process control systems, variables such as temperature, pressure, and flow replace the mechanical and electrical variables discussed in this chapter. The form of the transfer functions, however, will be the same. The processes will involve inertia and damping which will result in time constants and thus in transfer function poles.

2.15 SYSTEM BLOCK DIAGRAMS

All types of system can be conveniently represented by block diagrams. These can then be manipulated, using the transfer functions of the blocks, to obtain the system transfer function. A simple case, involving summing junctions and feedback loops will be illustrated in a final example.

Example 2.12. Fig. 2.15 shows the block diagram for a closed loop control system including negative velocity feedback and integral control. Determine the system transfer function.

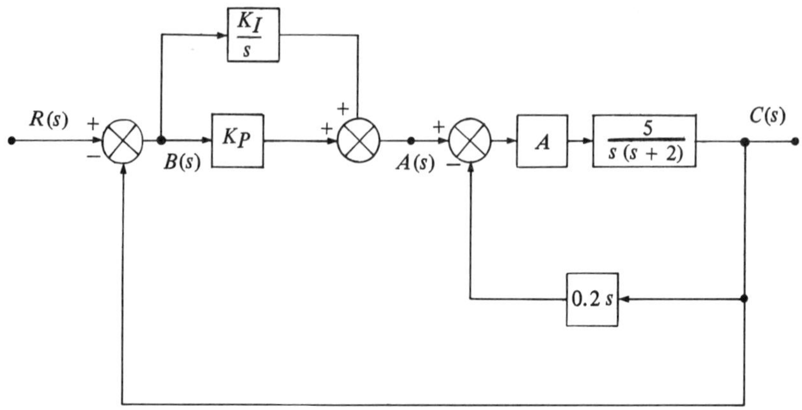

Fig. 2.15 The block diagram for a closed-loop control system.

Solution. Consider first the inner loop including the amplifier, the system being controlled and the negative velocity feedback 0.2*s*. Using the familiar feedback equation, the transfer function for this section is given by:

$$\frac{C}{D} = \frac{\dfrac{5A}{s(s+2)}}{1 + \dfrac{A}{(s+2)}} = \frac{5A}{s(s+A+2)} . \qquad \ldots (2.94)$$

The two parallel blocks in the forward path can be added and the result multiplied by (2.94).

$$\frac{C}{B} = \left(\frac{K_I}{s} + K_p\right) \times \frac{5A}{s(s+A+2)} . \qquad \ldots (2.95)$$

This result is rearranged and put into the unity feedback loop to give the system transfer function.

$$T(s) = \frac{C}{R} = \frac{\dfrac{5A(sK_p + K_I)}{s^2(s+2+A)}}{1 + \dfrac{5A(sK_p + K_I)}{s^2(s+2+A)}} ,$$

$$= \frac{5A(sK_p + K_I)}{s^2(s+2+A) + 5A(sK_p + K_I)} . \qquad \ldots (2.96)$$

This result can only be reduced further if values for the constants and amplifier gain are known. The system has a single zero and three poles which could be found by factorizing the cubic denominator. These poles may be real, or in complex conjugate pairs, depending on the value of *A*. Factorization of this type of denominator is discussed in Chapter 6.

This chapter opened with a definition of the term transfer function. Some applications of transfer functions were discussed briefly and the various forms of transfer function were described. The special component parts or parameters of the transfer function were then introduced as the poles and zeros of the transfer function. Calculation of the poles and zeros was then illustrated by an example. The remainder of the chapter was devoted to the techniques required to determine the transfer functions of a range of circuits and systems. These included passive electrical circuits, active circuits using isolating and operational amplifiers, simple mechanical systems and, briefly, some electro-mechanical systems involving d.c. motors. It can be stressed once again, that the general form of the transfer functions is the same in all cases. System transfer functions only differ in the units and in typical values for the poles and zeros or time constants.

2.16 EXAMPLES FOR FURTHER PRACTICE

Example 2.13. Express the following transfer functions in the Bode/time constant form.

a) $$\frac{(s+5000)\,(s+8000)}{(s+800)\,(s+50\,000)}\,.$$

b) $$\frac{5000\,(s^2+40s+300)}{s\,(s^3+500s^2+5000s)}\,.$$

c) $$\frac{s\,(s+5)\,(s^2+10s+9)}{(s+2)^3\,(s^2+4s+4)}\,.$$

Example 2.14. Express the following transfer functions in the pole zero form.

a) $$\frac{0.4}{s\,(1+0.05s)\,(1+0.01s)}\,.$$

b) $$\frac{100\,(1+10^{-4}s)}{(1+10^{-3}s)\,(1+5\times 10^{-6}s)}\,.$$

Example 2.15. Determine the poles and zeros for the transfer functions given in Examples 2.13 and 2.14 above and for the following transfer functions.

a) $$\frac{s\,(s^2-10^4)}{(s^2+10^4)}\,.$$

b) $$\frac{10}{s^2\,(s^2+2s+10)}\,.$$

c) $$\frac{(s^2+600s+10^5)}{(s^2+200s+2.6\times 10^5)^2}\,.$$

d) $$\frac{(s^2 + 1.5s - 1)}{(s^2 - 0.2s + 0.1)}.$$

(–800, –50 000; –5000, –8000: 0, 0, –10.2, –489.8; –10, –30: –2, –2, –2; 0, –5, –1, –9: 0, –20, –100: –1000, -2×10^5; -10^4: –j100, +j100, 0, –100, +100: 0, 0, –1 + j3, –1 – j3: –100 + j500, –100 – j500, –100 – j500, –100 – j500; –300 + j100, –300 – j100; +0.1 + j0.3, +0.1 – j0.3; +0.5, –2.0.)

Example 2.16. For each of the circuit diagrams shown in Fig. 2.16, determine the transfer function in terms of the input and output variables shown.

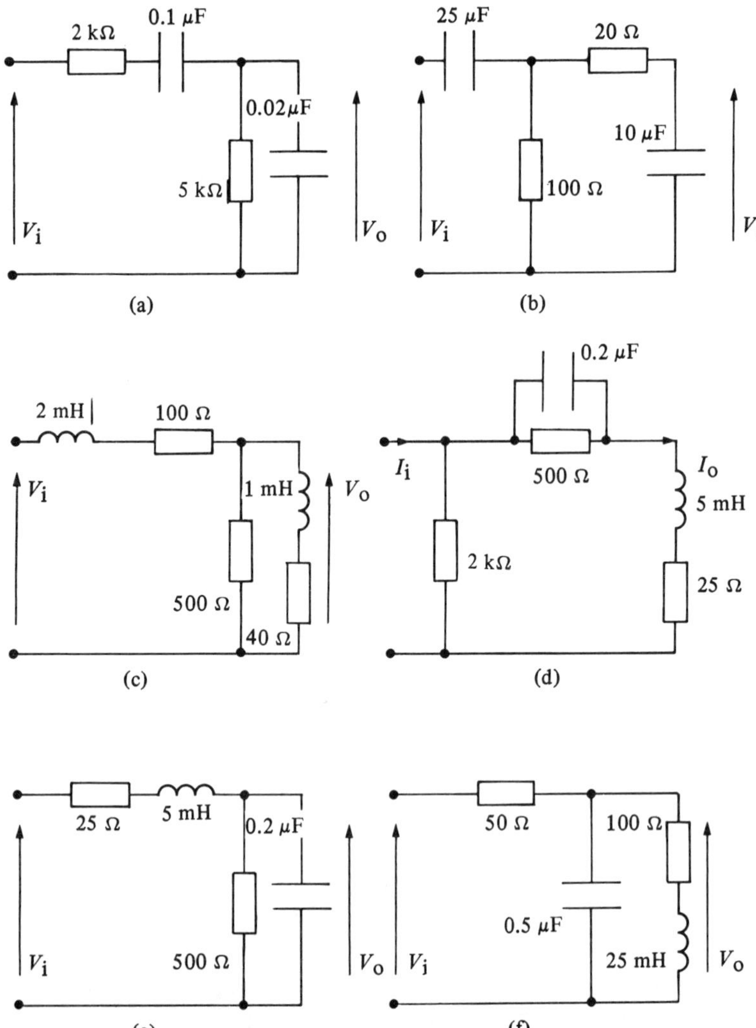

Fig. 2.16 Circuits for Example 2.16.

a) $$\frac{25\,000s}{(s + 1292)(s + 38\,708)}.$$

b) $$\frac{5000s}{(s + 281)(s + 7119)}.$$

c) $$\frac{25 \times 10^4 (s + 40\,000)}{(s + 4.7 \times 10^4)(s + 8.0 \times 10^5)}.$$

d) $$\frac{4 \times 10^5 (s + 10^4)}{(s + 1.25 \times 10^4)(s + 4.02 \times 10^5)}.$$

e) $$\frac{5 \times 10^9}{(s + 7500 - j3.2 \times 10^4)(s + 7500 + j3.2 \times 10^4)}.$$

f) $$\frac{40\,000 (s + 4000)}{(s^2 + 44\,000s + 2.4 \times 10^8)}.$$

Example 2.17. Determine, by application of mesh analysis, the voltage transfer function for the electrical network shown in Fig. 2.17.

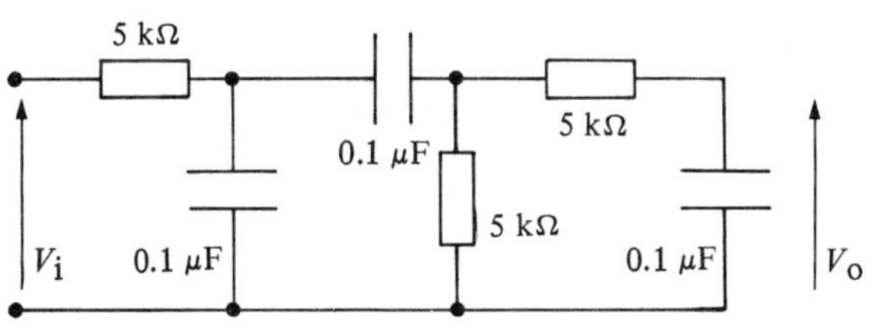

Fig. 2.17 Circuit for Example 2.17.

$$\frac{8 \times 10^6 s}{(s^3 + 12\,000s^2 + 2 \times 10^7 s + 8 \times 10^9)}$$

Example 2.18. Determine, by application of nodal analysis, the voltage transfer function for the scaled electrical network shown in Fig. 2.18.

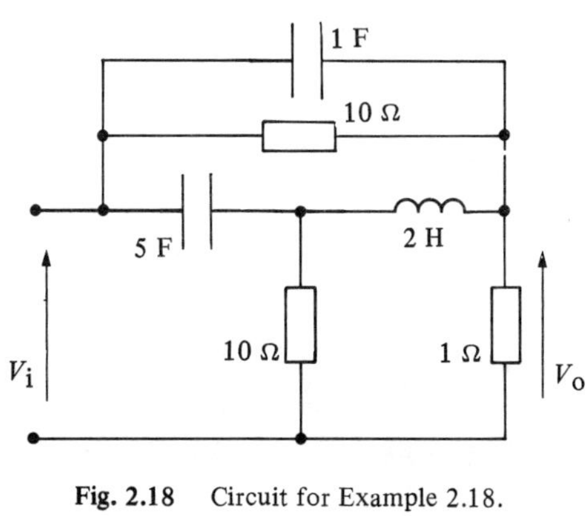

Fig. 2.18 Circuit for Example 2.18.

$$\frac{s^3 + 0.15s^2 + 0.102s + 0.01}{s^3 + 1.12s^2 + 0.62s + 0.12}.$$

Example 2.19. The feedback circuit shown in Fig. 2.19 is that of a selective bandpass amplifier. Assuming the amplifier to be 'ideal', determine the system transfer function.

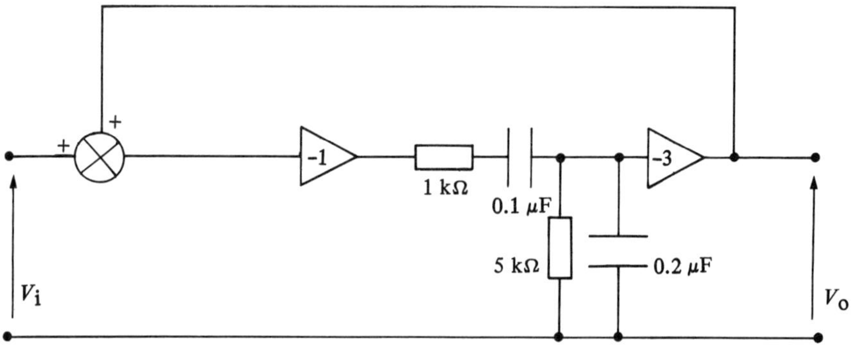

Fig. 2.19 Feedback circuit for Example 2.19.

$$\frac{15\,000s}{(s + 500 - j3122)\,(s + 500 + j3122)}$$

Example 2.20. Determine the voltage transfer functions for the two operational amplifier circuits shown in Fig. 2.20.

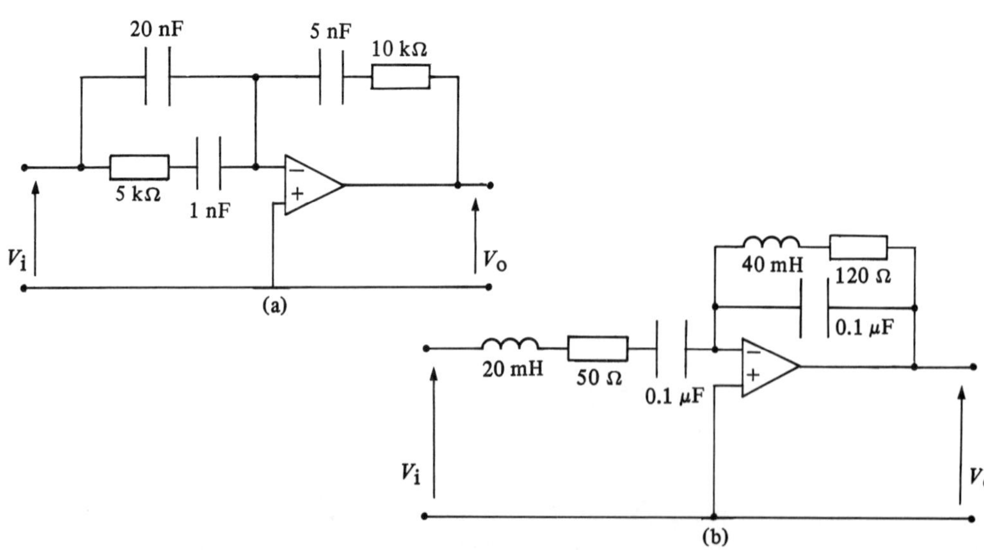

Fig. 2.20 Operational amplifier circuits for Example 2.20.

a) $$\frac{2 \times 10^{-4}\,(s + 2.1 \times 10^5)\,(s + 2 \times 10^4)}{(s + 2 \times 10^5)},$$

b) $$\frac{5 \times 10^8\,(s + 3000)s}{(s^2 + 2500s + 5 \times 10^8)\,(s^2 + 3000s + 2.5 \times 10^8)}.$$

Example 2.21. Fig. 2.21 shows a system constructed from an operational amplifier and two three-terminal networks. The circuits for the three-terminal networks are shown respectively in Fig. 2.21b and c. Determine the system transfer function.

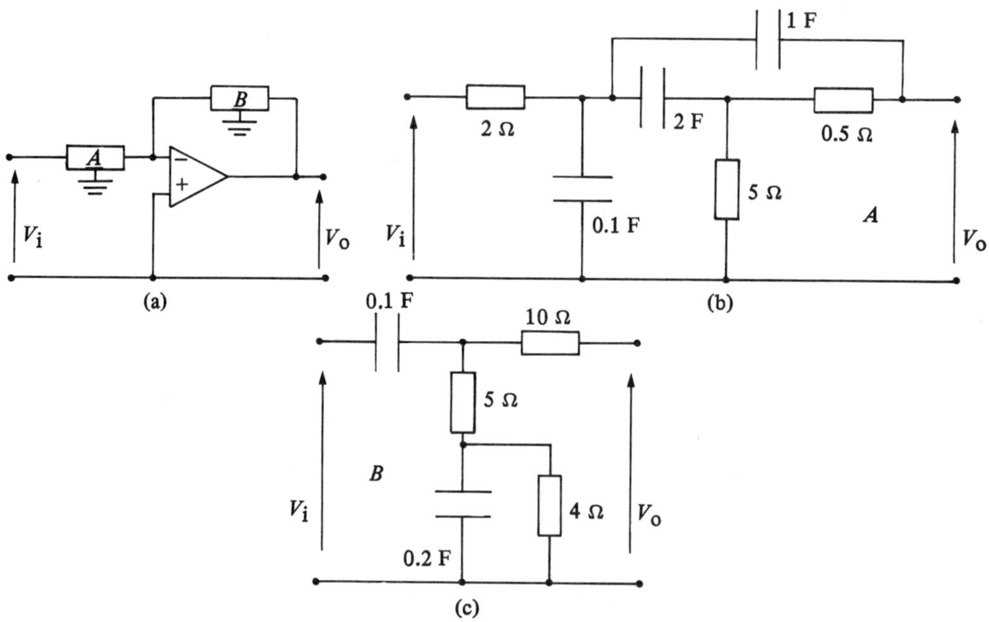

Fig. 2.21 Operational amplifier circuit for Example 2.21.

$$\frac{5s\,(s + 3.1)\,(s + 0.198)\,(s + 5.05)}{(s^3 + 3.8s^2 + 10.55s + 11)\,(s + 2.25)}.$$

Example 2.22. Two active filter circuits are shown in Fig. 2.22. With appropriate values they may be employed respectively as bandpass and low pass filters. In each case, determine the voltage transfer function.

a) $$\frac{-s}{s^2 + s\left(\dfrac{1}{C_2R_2} + \dfrac{1}{C_1R_2} + \dfrac{2}{C_1R_1}\right) + \dfrac{1}{C_1C_2R_1R_2}}.$$

b) $$\frac{-G_1G_3}{s^2C_2C_5 + sC_5(G_1 + G_3 + G_4) + G_3G_4}.$$

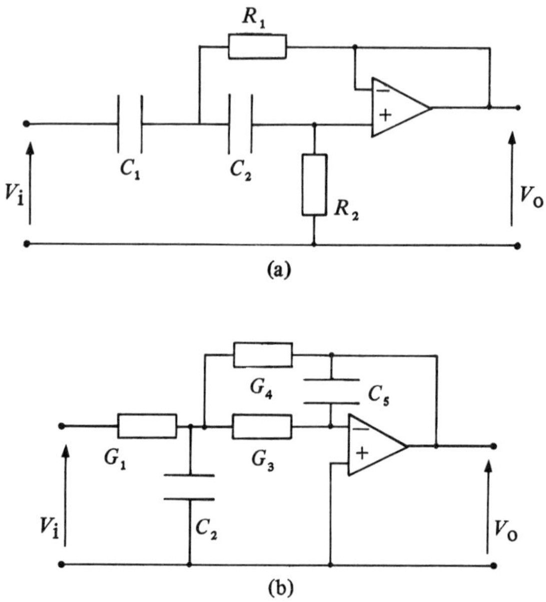

Fig. 2.22 Active filter circuits for Example 2.22.

Example 2.23. A field controlled d.c. motor is used to control the angular position of a rotatable load having inertia 0.02 kg m^2 and viscous friction 0.14 Nm rads^{-1}. The motor torque, for a particular armature current setting, is 10^{-4} Nm rads^{-1} for a field current of 1 mA. The field circuit resistance and inductance are 5 Ω and 0.1 H respectively. The constants for the combined transducers and differential result in 2 V per radian of error between input and output angles. This voltage is amplified by an amplifier having a gain of 50, the resulting output being applied to the motor field circuit. Determine the system transfer function.

$$\left(\frac{5000}{s^3 + 57s^2 + 350s + 5000}\right)$$

Example 2.24. The block diagram shown in Fig. 2.23 represents a multiple-loop control system with controlled output C and reference input R. Determine the system transfer function $\frac{C}{R}(s)$.

$$\left(\frac{0.5\,(s^3 + 3.2s^2 + 4.4s + 16)}{(s^3 + 6.2s^2 + 7s + 8)}\right).$$

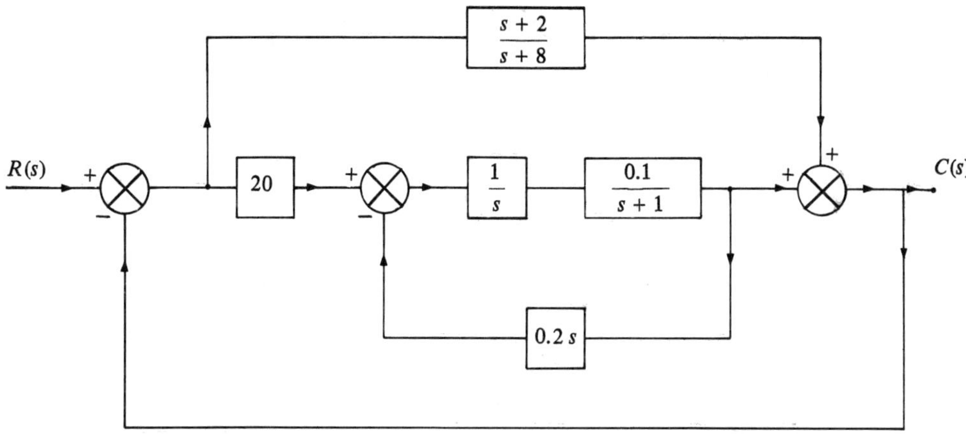

Fig. 2.23 Block diagram for Example 2.24.

3

The Pole Zero Diagram and the Transient Response

3.1 INTRODUCTION

This chapter is concerned with the pole zero diagram and the relationship between this diagram and the system transient response. The representation of transfer and response functions by means of a pole zero diagram is introduced and the details of such diagrams are explained. The relationship between the diagram and the step response is then shown for both overdamped and underdamped systems. This approach is then extended to cover the response of systems to other switched input signals. Finally, a number of general relationships are illustrated and these are then used to obtain a set of useful guidelines.

3.2 THE POLE ZERO DIAGRAM

In Chapter 2, we have seen how a transfer function is characterized by the critical values of complex frequency known as poles and zeros. It is often convenient to describe or illustrate such a transfer function by means of a pole zero diagram which shows the positions of these critical frequencies plotted in the complex plane. As an example, consider the following transfer function:

$$T(s) = \frac{500(s+2)}{s\,(s+5)\,(s^2+2s+17)} \,. \qquad \ldots(3.1)$$

This transfer function has a zero at $s = -2$ and poles at $s = 0, -5, -1 + j4$ and $-1 - j4$. The pole zero diagram for this function is shown in Fig. 3.1. The following points should be observed:

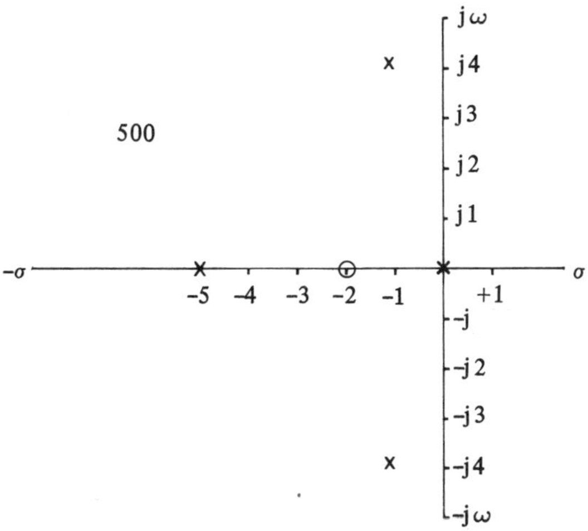

Fig. 3.1 The pole zero diagram for $T(s) = \dfrac{500\,(s+2)}{s\,(s+5)\,(s^2+2s+17)}$.

i. The axes are $\pm\sigma$ and $\pm j\omega$ and they are both drawn to the same linear scale.
ii. A zero is a value of s which makes $T(s) = 0$. In this case, the value $s = -2$ will make $T(s)$ or $T(-2) = 0$ The point -2 on the real axis is shown as a zero.
iii. A pole is a value of s which makes $T(s) = \infty$. In this case there are four such values at: $s = 0$, $s = -5$ and the values of s which make $s^2 + 2s + 17 = 0$. These four poles are plotted on the pole zero diagram as X.
iv. The multiplying or scaling factor of 500 is written on the diagram for convenience.

At this stage, it should be noted that the diagram contains no more information than the original transfer function. On the other hand, a map of a district is easier to appreciate than a list of map references. In a similar way, the pole zero diagram can be regarded as a map of the transfer function. The transient and steady-state properties of the system producing the transfer function can be estimated, or if required, determined precisely from the pole zero diagram. In the next section we see how the pole zero diagram can help in the calculation of the transient response (usually to a step input). This will then lead to general conclusions about the transient response resulting from particular pole zero arrangements.

3.3 THE POLE ZERO DIAGRAM AND THE STEP RESPONSE

In general,

$$T(s) = \frac{\text{output}}{\text{input}}(s)\ . \qquad \ldots(3.2)$$

In control engineering, the output and input are the controlled variable $C(s)$ and the reference input $R(s)$ respectively. This results in the transfer function form,

$$T(s) = \frac{C(s)}{R(s)}\ . \qquad \ldots(3.3)$$

However, the use of C for the output of an electrical or other system is not appropriate as the element of control does not apply. In these more general cases, the output is the *response* to a driving or exciting signal. We shall therefore, in this context, use $R_{(s)}$ and $D(s)$ for the response and input respectively. The resulting transfer function form is,

$$T(s) = \frac{R(s)}{D(s)} . \qquad \ldots(3.4)$$

Thus the s domain response will be given by, $R(s) = T(s) \times D(s)$, and the resulting time domain response will be $r(t)$.

For the step response, $D(s)$ is the transform for a step of amplitude a which is simply a/s or more often $1/s$ if the input is a unit step. Thus to obtain the step response, we simply multiply $T(s)$ by $1/s$. The inverse transform procedure must then be used to find the time domain response $r(t)$.

Example 3.1. Determine the step response for a system having the following transfer function:

$$T(s) = \frac{192\,(s+2)}{(s+4)\,(s+8)} .$$

Solution.

$$R(s) = \frac{1}{s} \times T(s) = \frac{192\,(s+2)}{s\,(s+4)\,(s+8)} . \qquad \ldots(3.5)$$

For the partial fraction expansion, equate expression (3.5) to the required form,

$$\frac{192\,(s+2)}{s\,(s+4)\,(s+8)} = \frac{A}{s} + \frac{B}{s+4} + \frac{C}{s+8} , \qquad \ldots(3.6)$$

where the ***residues*** A, B and C are found by the cover up rule.

$$A = \frac{192 \times (+2)}{(+4) \times (+8)} = 12 . \qquad \ldots(3.7)$$

$$B = \frac{192 \times (-2)}{(-4) \times (+4)} = 24 . \qquad \ldots(3.8)$$

$$C = \frac{192 \times (-6)}{(-8) \times (-4)} = -36 . \qquad \ldots(3.9)$$

Reference to the transform tables shows that the time domain response is given by:

$$r(t) = 12 + 24e^{-4t} - 36e^{-8t} . \qquad \ldots(3.10)$$

Let us now compare this result with the pole zero diagram shown in Fig. 3.2a. This diagram is strictly that for the response as it includes the pole at the origin resulting from the input step transform $1/s$. Notice that each pole results in a term in the response $r_{(t)}$ and that the time constants of the exponentials are the reciprocals of the pole frequencies. i.e.

$$r(t) = 12 + 24e^{-t/0.25} - 36e^{-t/0.125} \quad \ldots(3.11)$$

The zero at $s = -2$ does not result in a response term but contributes to the amplitude and sign of the transient terms resulting from the poles. A graph of the resulting time domain response $r(t)$ is shown in Fig. 3.2b.

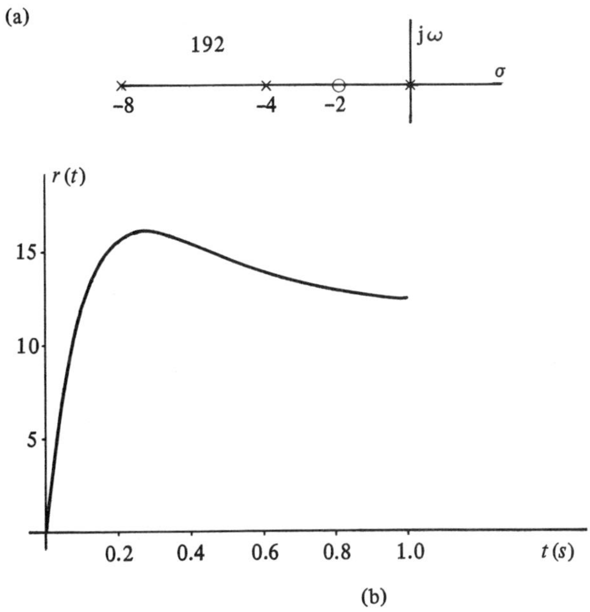

Fig. 3.2 a) The response pole zero diagram for Example 3.1. b) The time domain step response for Example 3.1.

3.4 DETERMINATION OF THE PARTIAL FRACTION RESIDUES FROM THE POLE ZERO DIAGRAM

It will now be useful to examine the partial fraction residue expressions obtained by the cover up rule and compare them with the pole zero diagram as shown in Fig. 3.3.

It can be seen that each term in brackets in the equations (3.7) to (3.9) is the length of line or segment drawn between a pole and a zero or between two poles. These *line segments* are shown *directed towards* the pole in question (i.e. Fig. 3.3b for the residue B at the pole at $s = -4$ shows all the line segments directed towards the pole at $s = -4$). Finally, the sign of each term in brackets corresponds to the direction of the line segment; those pointing to the right are positive and those to the left are negative.

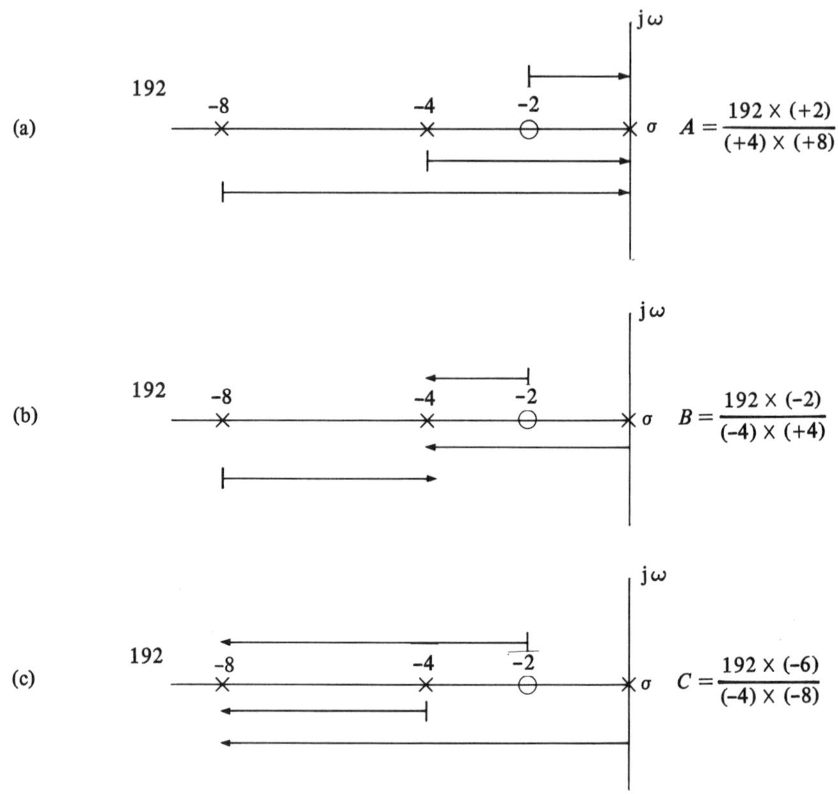

Fig. 3.3 The relationship between partial fraction residues and the pole zero diagram.

The above relationships suggest a general rule for the determination of the residue A at a simple pole of a function shown on a pole zero diagram.

$$A = K\left[\frac{\text{The product of the directed line segments } \textit{from} \text{ each zero } \textit{to} \text{ the pole in question.}}{\text{The product of the directed line segments } \textit{from} \text{ each other pole } \textit{to} \text{ the pole in question.}}\right] \qquad \ldots(3.12)$$

Note that in each case, the direction is *towards* the pole in question, that is, the pole for which the residue is being determined. A simple pole is one that occurs when there are not two or more equal roots resulting in coincident poles.

In Example 3.1, there is no great advantage in the use of this rule to find the residues, as only real poles and zeros are present. However, transfer functions may include complex poles and zeros, when a considerable advantage will be obtained by the use of this method to find the residues. The general expression that is derived in the next section can then be used to obtain the resulting transient terms.

3.5 GENERAL EXPRESSION FOR TRANSIENT RESPONSE TERMS RESULTING FROM COMPLEX CONJUGATE PAIRS OF POLES

Consider a transfer function having a number of poles and zeros including a complex conjugate pair of poles at:

$$s_1 = -\sigma + j\omega \quad \text{and} \quad s_2 = -\sigma - j\omega. \qquad \ldots(3.13)$$

The residues at s_1 and s_2 can be found, either by the cover up rule or by the directed line segment method described in the last section. These residues will also be complex and conjugate and may therefore be written:

$$B = a + jb \quad \text{and} \quad B^* = a - jb\ . \qquad \ldots(3.14)$$

Thus, the part of the response resulting from these terms will be:

$$R(s) = \frac{a + jb}{s + \sigma - j\omega} + \frac{a - jb}{s + \sigma + j\omega}\ . \qquad \ldots(3.15)$$

Applying the inverse transform $\left(\dfrac{K}{s + d} \rightarrow K e^{-dt}\right)$,

$$r(t) = (a + jb)\, e^{-(\sigma - j\omega)t} + (a - jb)\, e^{-(\sigma + j\omega)t}\ . \qquad \ldots(3.16)$$

Now, extracting the common factor $e^{-\sigma t}$ and collecting terms,

$$r(t) = e^{-\sigma t}\left[a\,(e^{j\omega t} + e^{-j\omega t}) + jb\,(e^{j\omega t} - e^{-j\omega t})\right]. \qquad \ldots(3.17)$$

But the exponential forms for $\cos\theta$ and $\sin\theta$ are:

$$\cos\theta = \frac{(e^{j\theta} + e^{-j\theta})}{2} \quad \text{and} \quad \sin\theta = \frac{(e^{j\theta} - e^{-j\theta})}{2j}\ . \qquad \ldots(3.18)$$

Rearranging $r(t)$ to match these forms,

$$r(t) = 2e^{-\sigma t}\left[a\,\frac{(e^{j\omega t} + e^{-j\omega t})}{2} - b\,\frac{(e^{j\omega t} - e^{-j\omega t})}{2j}\right]. \qquad \ldots(3.19)$$

$$\therefore \quad r(t) = 2e^{-\sigma t}\,(a \cos\omega t - b \sin\omega t). \qquad \ldots(3.20)$$

This result can now be compared with the identity:

$$K \cos(\omega t + \theta) = K \cos\omega t \cos\theta - K \sin\omega t \sin\theta\ . \qquad \ldots(3.21)$$

Hence,

$$a = K\cos\theta \quad \text{and} \quad b = K\sin\theta;$$

and since

$$a^2 + b^2 = K^2\,(\cos^2\theta + \sin^2\theta) = K^2,$$

then,

$$K = \sqrt{a^2 + b^2} \ . \qquad \ldots(3.22)$$

Reference to the original definition for B in expression (3.14) shows that:

$$K = |B| \ . \qquad \ldots(3.23)$$

Also,

$$\frac{b}{a} = \frac{K \sin\theta}{K \cos\theta} = \tan\theta \ .$$

$\therefore \ \theta = \tan^{-1} b/a$ which from expression (3.14) is the same as $\angle B$. ...(3.24)

Combining these results, the response $r(t)$ may be rewritten:

$$r(t) = 2|B|\, e^{-\sigma t} \cos(\omega t + \angle B) \ ; \qquad \ldots(3.25)$$

or, in a more usual form,

$$r_{(t)} = |2jB|\, e^{-\sigma t} \sin(\omega t + \angle 2jB) \ . \qquad \ldots(3.26)$$

It should be noted in this expression that the j has no effect upon $|2jB|$ and that the 2 has no effect upon $\angle 2jB$ which is simply $\underline{/B + 90^\circ}$. The application of the directed line segment method and this general result can now be illustrated by some examples.

Example 3.2. Determine the step response for systems having the following transfer functions.

a) $$T_1(s) = \frac{58}{(s^2 + 4s + 29)} \ . \qquad \ldots(3.27)$$

b) $$T_2(s) = \frac{10^6(s + 200)}{(s + 500)(s^2 + 100s + 362\,500)} \ . \qquad \ldots(3.28)$$

c) $$T_3(s) = \frac{20s(s^2 + 6s + 18)}{(s + 2)(s^2 + 2s + 10)} \ . \qquad \ldots(3.29)$$

Solution. a) After factorizing the quadratic term, the step response for $T_1(s)$ is given by:

$$R(s) = \frac{58}{s(s + 2 - j5)(s + 2 + j5)} = \frac{A}{s} + \frac{B}{s + 2 - j5} + \frac{B^*}{s + 2 + j5} \ . \qquad \ldots(3.30)$$

The residues A and B (B^* is not required for the general form (3.26)) can now be found by the directed line segment method (3.12). The required response pole zero diagram is shown in Figs. 3.4a and 3.4b. In Fig. 3.4a, the directed line segments required to find residue A are shown. Those required for B are shown in Fig. 3.4b.

Applying expression (3.12) for the determination of residues;

$$A = \frac{58}{\sqrt{(2^2 + 5^2)}\,\underline{/\tan^{-1} 5/2} \times \sqrt{(2^2 + 5^2)}\,\underline{/\tan^{-1} -5/2}} = 2\,\underline{/0^\circ} \ . \qquad \ldots(3.31)$$

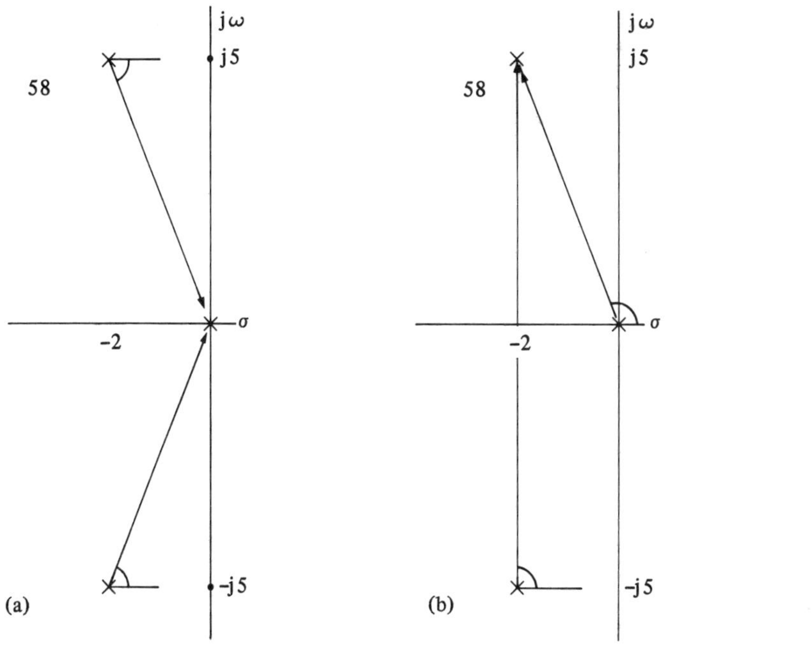

Fig. 3.4 The response pole zero diagram and directed line segments for Example 3.2a.

Notice that the two angles cancel; this will always apply when the residue of a pole on the real axis is being determined. When there are other poles or zeros on the real axis, the total angle will be either 0° or 180° which is then usually expressed as a minus sign. Also note that the two lengths are equal, simplifying the result to $\dfrac{58}{(2^2+5^2)}$.

Applying expression (3.12) to determine residue B,

$$B = \frac{58}{\sqrt{(2^2+5^2)}\angle\underline{180 - \tan^{-1} 5/2} \times 10\angle 90}\ ; \qquad \ldots(3.32)$$

$$B = \frac{1.077\angle\underline{-111.8^\circ}}{1\angle 90^\circ}\ .$$

In this result, the 90° from the conjugate pole has been left in the denominator; this is because it will cancel with the j in result (3.26). The general expression, (3.26), is given in terms of $|2jB|$ and $\angle 2jB$; in this example, $2jB = 2.154\angle\underline{-111.8^\circ}$ and the response can be written from (3.30) as:

$$r(t) = 2 + 2.154e^{-2t}\sin(5t - 111.8^\circ)\ . \qquad \ldots(3.33)$$

It is a useful check to note that substituting $t = 0$ into this result shows:

$$r(o) = 2 + 2.154\sin(-111.8^\circ) = 0.$$

A zero result is not universal as some transfer functions will produce a step in the response. This aspect is discussed later in this chapter.

b) The step response for this transfer function is given by:

$$R(s) = \frac{10^6\,(s+200)}{s\,(s+500)\,(s+50-\mathrm{j}600)\,(s+50+\mathrm{j}600)}\,,$$

and for partial fraction expansion this is equated to:

$$R(s) = \frac{A}{s} + \frac{B}{(s+500)} + \frac{C}{(s+50-\mathrm{j}600)} + \frac{C^*}{(s+50+\mathrm{j}600)}\,. \qquad \ldots(3.34)$$

The pole zero diagram for this response is shown in Fig. 3.5.

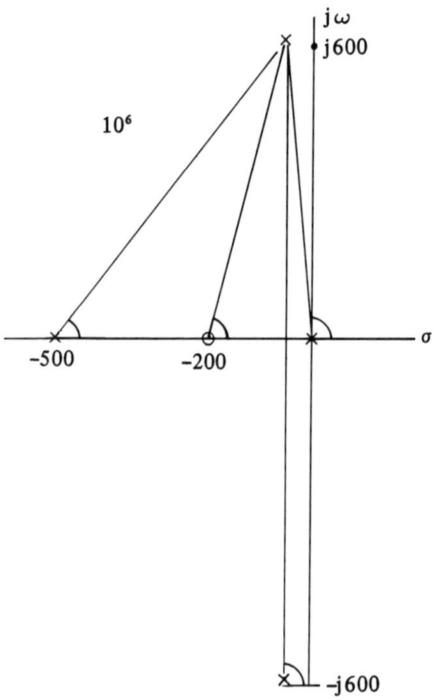

Fig. 3.5 The response pole zero diagram and directed line segments for Example 3.2b.

The directed line segments for the residues A and B are found in the same way as for the previous transfer function; the segments for C are shown so that the correct angles can be determined.

$$A = \frac{10^6 \times 200}{500 \times (600^2 + 50^2)} = +1.1\,. \qquad \ldots(3.35)$$

$$B = \frac{10^6 \times 300}{500 \times (600^2 + 450^2)} = +1.067\,. \qquad \ldots(3.36)$$

Note that both A and B are positive. This is because in each case, there is an even number of poles and zeros to the right of the pole in question. Each such pole and zero contributes 180° or a minus sign, an even number will therefore result in a positive term.

$$C = \frac{10^6 \times \sqrt{(600^2 + 150^2)} \angle \tan^{-1}\frac{600}{150}}{\sqrt{(600^2 + 50^2)} \angle 180 - \tan^{-1}\frac{600}{50} \times \sqrt{(600^2 + 450^2)} \angle \tan^{-1}\frac{600}{450} \times 1200 \angle 90^\circ} . \qquad \ldots (3.37)$$

$$C = \frac{1.14 \angle -72^\circ}{\angle 90^\circ}$$

Hence,

$$r(t) = 1.1 + 1.067e^{-500t} + 2.28e^{-50t} \sin(600t - 72^\circ) . \qquad \ldots (3.38)$$

and once again,

$$r(o) = 0 .$$

c) In this case, the input step cancels the transfer function zero at the origin and there are complex zeros to take into account.

$$R(s) = \frac{20(s+3-j3)(s+3+j3)}{(s+2)(s+1-j3)(s+1+j3)} = \frac{A}{s+2} + \frac{B}{s+1-j3} + \frac{B^*}{s+1+j3} \qquad \ldots (3.39)$$

The response pole zero diagram is shown in Fig. 3.6 and, once again, the directed line segments are only shown for the residue B at the complex conjugate pole.

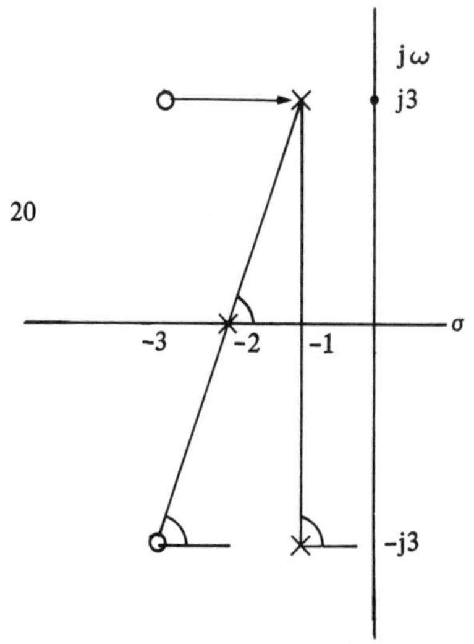

Fig. 3.6 The response pole zero diagram and directed line segments for Example 3.2c.

It is left to the reader to determine the residues and then to find that the response $r(t)$ is given by,

$$r(t) = 20e^{-2t} + 26.6e^{-t} \sin 3t \ . \qquad \ldots (3.40)$$

For this example, $r(o)$ is not zero.

3.6 STEP RESPONSES FOR PASSIVE ELECTRICAL CIRCUITS

This section will be used to show a number of circuits with, in each case, the pole zero diagram for the voltage transfer function and a sketch of the resulting step response. This will enable us to find a number of general rules which will allow us to 'read' the pole zero diagram and to estimate the probable form of the step response without calculation. Such results would, of course, apply to non-electrical systems having similar pole zero configurations.

3.7 CIRCUITS WITH NO ZEROS AND ONLY REAL POLES

The first group of circuits are shown in Fig. 3.7. These are *RC* ladder networks with one, two and three sections. In each case, the product *CR* is chosen to be 10^{-3} s and the resulting response pole zero diagram is shown including the pole for the input step. It can be seen how the addition of the second and third section respectively, interact with the original circuit pole and move it towards the origin. The choice of $C/2$ and $C/4$ for the second and third sections reduces the resulting 'spread' of the poles.

This first point to note from the pole zero diagrams is that each addition of a pole 'pushes' the other poles towards the origin. Next, referring to the expressions for $r(t)$, we can see that the residues alternate in sign. The reason for this can be seen in the general result (3.12). Each pole or zero to the right of the pole in question will introduce a negative line segment. Thus, an even number to the right produces a positive residue and an odd number to the right, a negative residue. From the same general result, poles near each other will result in small denominator terms and thus larger residues. Poles distant from other poles will, by the same reasoning, have smaller residues. In case (a), with only two response poles, the residues are of equal size. In (b), the first two poles are near to each other and have relatively large residues while the third is distant from both and has a smaller residue. This last pole, however, has a greater effect on the middle pole than on the first. The same effects are also obvious in case (c).

The resulting effects upon the time response are also clear. In every case, $r(t)$ starts at zero and finishes by approaching unity exponentially with a time constant due to the circuit pole nearest the origin. In circuit (b), the components due to the system poles tend to cancel for small values of time, resulting in a delay before the main exponential rise starts. Each additional pole increases this effect.

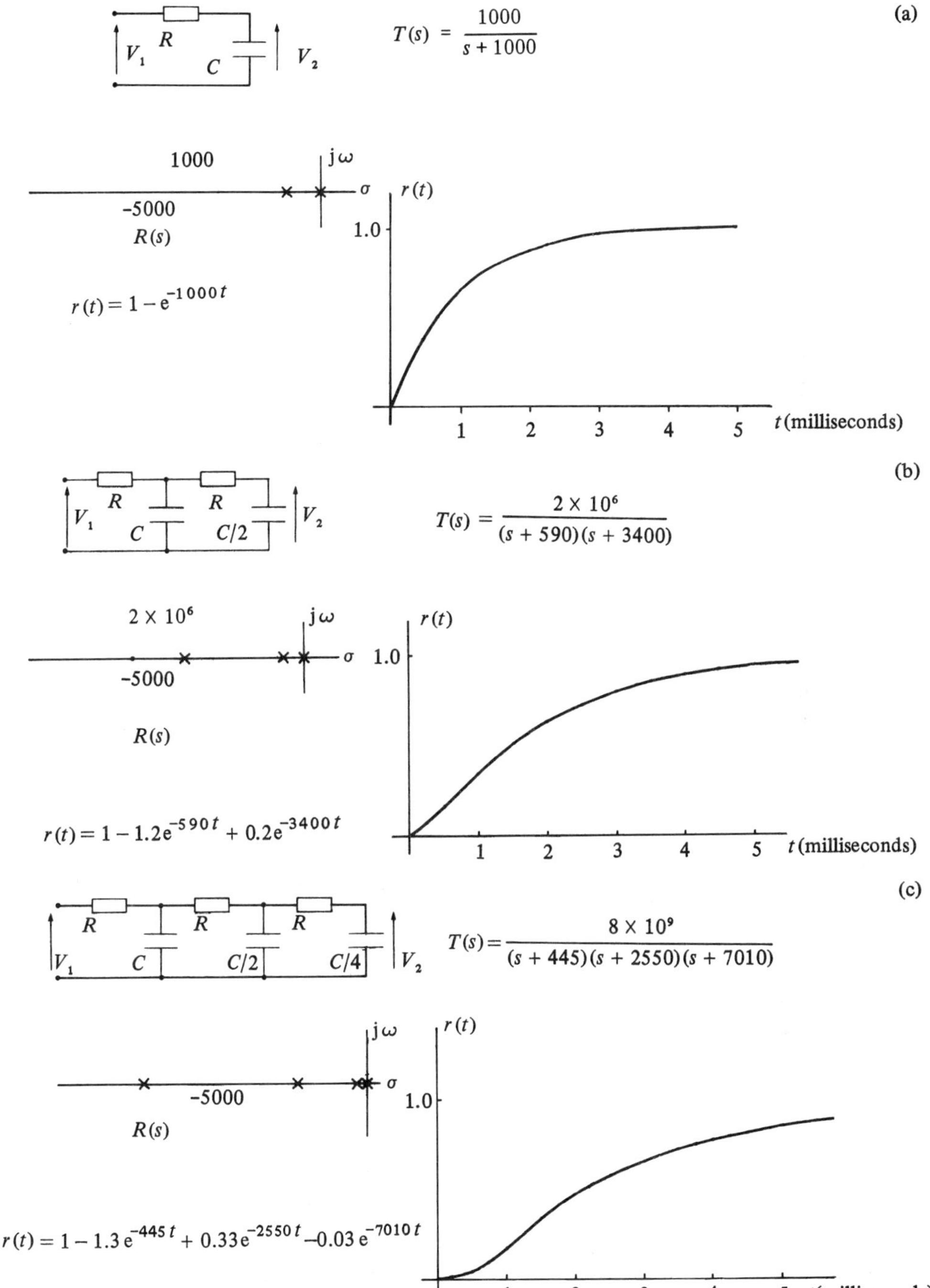

Fig. 3.7 The transient response of some low pass filter circuits having transfer functions with no zeros and only real poles.

The initial and final values of the response can also be deduced in the following way. The final value of the response is the d.c. value, for which $s = 0$. Thus if this value of s is substituted in the expression for $R(s)$, the steady-state component of $r(t)$ is obtained. For all the examples in this section, the result is unity, but the circuit shown in Fig. 3.8b in the next group is an example resulting in a finite non-unity steady-state response.

For the initial value, we are concerned effectively with the very high frequency response. We can therefore let s tend to infinity to obtain the response at $t = 0$. In each of the above cases, the result is obviously zero but in the following sections this approach will be applicable.* For example, with the circuit in Fig. 3.8a, as s tends to infinity, $T(s)$ tends to

$$0.2 \times \frac{s}{s} = 0.2.$$

3.8 CIRCUITS WITH REAL POLES AND REAL ZEROS

The next group of circuits shown in Fig. 3.8 consist of various RC networks, each of which have some finite real zeros as well as poles.

As with the first group, a standard CR value of 10^{-3} s has been used to help with making comparisons. The signs of the residues should again be noticed as the introduction of zeros now changes the previous pattern of alternating sign. The rule relating the odd or even number of poles and zeros to the right still applies. This is particularly obvious in cases (b), (c) and (d). The second important effect is the reduction of residue magnitude when a zero is near the pole in question. This can be clearly seen in cases (a) and (b). In (a), the zero is nearer to the system pole and the residue of this pole is therefore smaller than that of the step pole. In (b), the position is reversed and the step is reduced at the expense of the exponential term. In circuit (c), the residue of the pole at $s = -590$ is affected by the pole on one side and the zero on the other and is therefore larger than that at the pole at $s = -3600$ whose magnitude is dominated by the zero. The poles in (c) are the same as those in Fig. 3.7b but in this case, the zero between the poles reverses the sign of the residue at

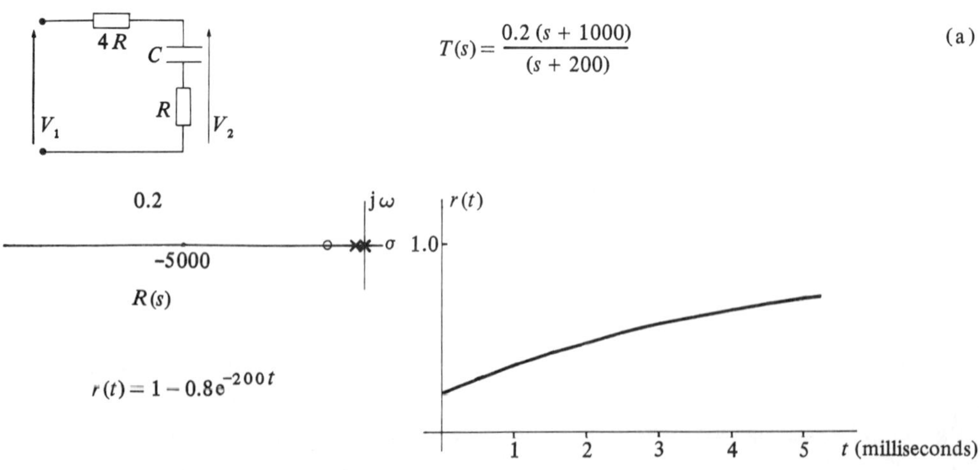

* See note on page 98.

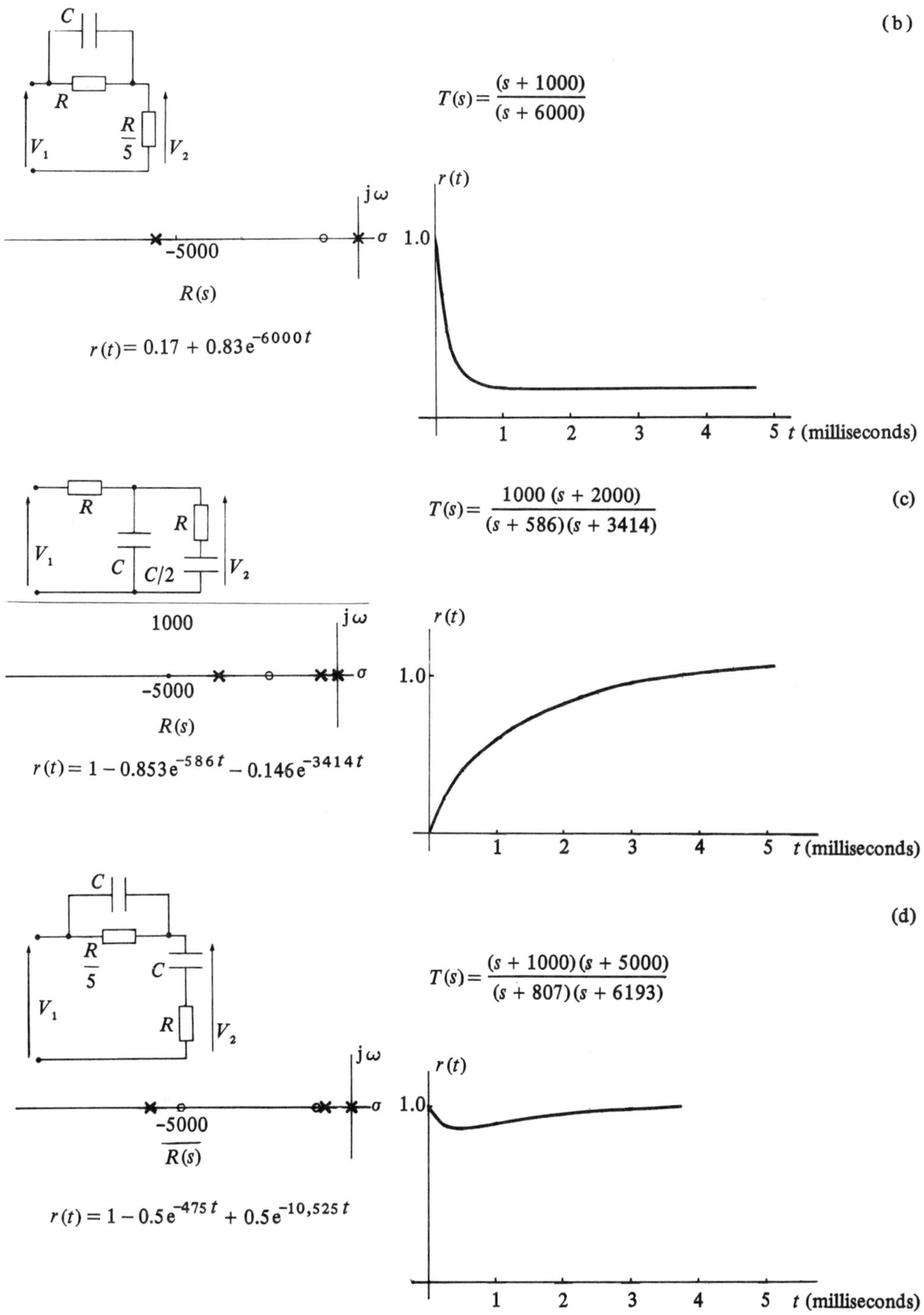

Fig. 3.8 The transient response of some *RC* compensator networks having transfer functions with real zeros and real poles.

the second pole. This results in the early part of the response being accelerated instead of delayed as in 3.7b. In circuit (d), the transfer function has a unity constant as the product of the zeros is equal to the product of the poles. This results in the response having equal amplitude exponents and a unity steady-state value.

If the product of the poles is not equal to that of the zeros, the transfer function will include a non-unity constant. The response will still include a unit step, but the amplitude of the exponents will no longer be equal.

3.9 CIRCUITS WITH REAL POLES AND ZEROS, INCLUDING ZEROS AT THE ORIGIN

The last group including only real poles and zeros are circuits having transfer functions with one or more zeros at the origin. This shows that the d.c. steady-state response will be zero, often the result of a series capacitor if the transfer function is a voltage ratio function. Fig. 3.9 shows some examples of this type.

In cases (a), (c) and (d), the input step pole is shown cancelling the zero of the transfer function. In case (b), there is a double zero, one of which is cancelled to leave a zero in the response function. In (a), the response is the result of a simple pole and is thus a single exponent. In (b), we have two poles, and a zero at the origin. This results in a response with the long time constant being negative and small compared with the positive short time constant term. The resultant therefore starts with a positive step and then crosses to a negative value before decaying to zero. In (c), the response function simply has two poles with the corresponding equal amplitude and opposite sign exponents.

The last circuit, (d), has the same poles as circuit (b) but one of the zeros at the origin has been shifted to a point between the poles. The result, in the time response, is to make both exponents positive. The zero is nearer to the low frequency pole and thus the long time constant term is smaller than the short time constant term resulting from the high frequency pole. The output step is therefore followed by a very steep initial fall followed by a long exponential decay to zero.

$$T(s) = \frac{s}{(s + 1000)} \qquad \text{(a)}$$

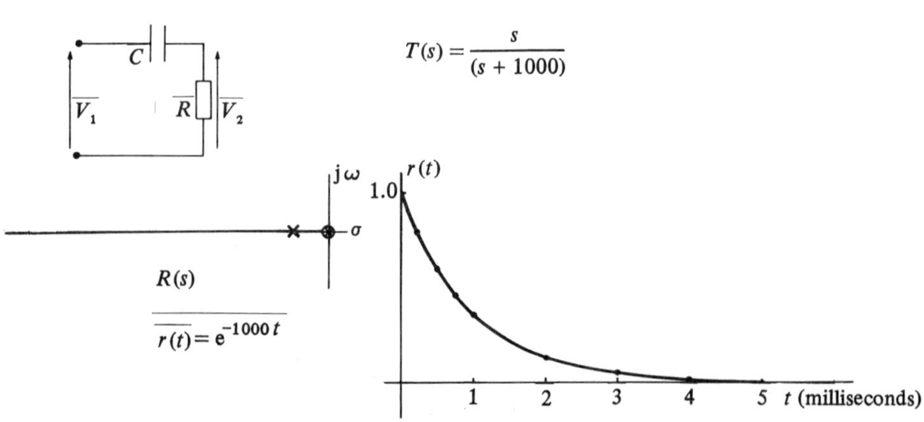

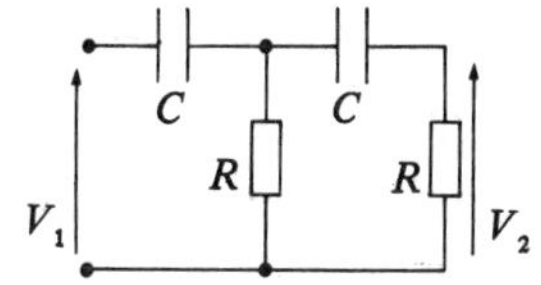

$$T(s) = \frac{s^2}{(s + 382)(s + 2618)} \qquad \text{(b)}$$

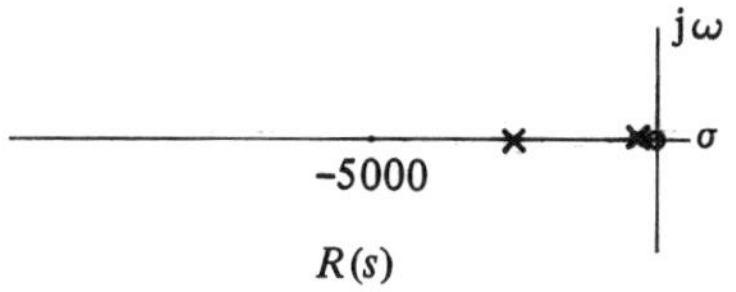

$$r(t) = 1.17e^{-2618t} - 0.17e^{-382t}$$

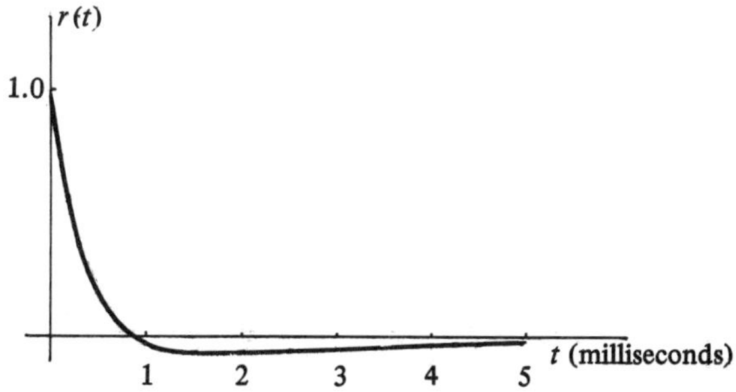

(c)

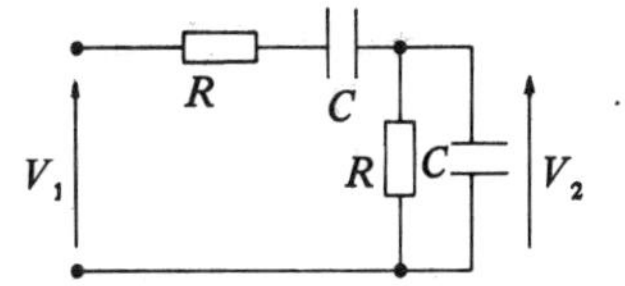

$$T(s) = \frac{1000s}{(s + 382)(s + 2618)}$$

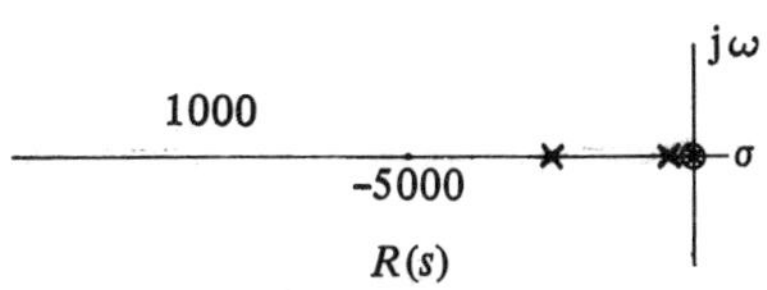

$$r(t) = 0.45e^{-382t} - 0.45e^{-2618t}$$

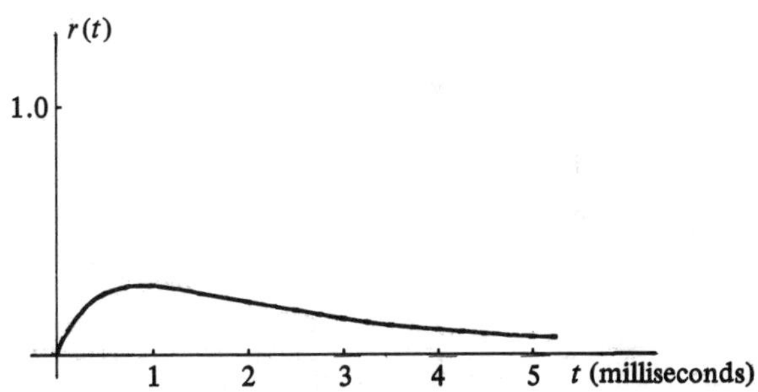

(d)

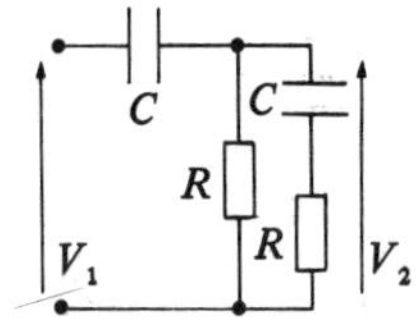

$$T(s) = \frac{s(s + 1000)}{(s + 382)(s + 2618)}$$

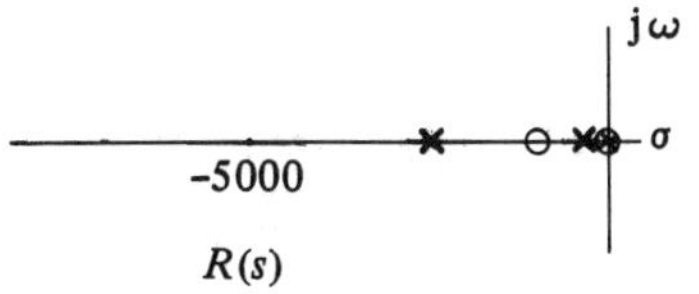

$$r(t) = 0.276e^{-382t} + 0.724e^{-2618t}$$

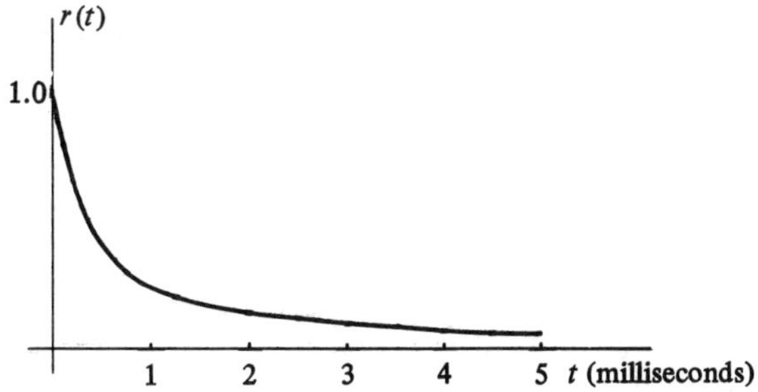

Fig. 3.9 The transient response of some high pass *CR* networks having transfer functions with real poles and zeros including a zero at the origin.

3.10 MULTIPLE REAL POLES

A problem that may be encountered is the case of double, real poles which result in a critically damped response. In this situation, the residues at any other poles can be found by the cover up rule or from the pole zero diagram, but we are left with an additional term which requires other methods. An alternative approach is to consider the double pole as two separate poles, very close to each other. The resulting response will be indistinguishable, as may be seen from the following example.

Example 3.3. Compare the step responses of the transfer functions;

$$T_1(s) = \frac{2000s}{(s+20)^2\,(s+40)}, \qquad \ldots(3.41)$$

and the approximate equivalent:

$$T_2(s) = \frac{2000s}{(s+19)\,(s+21)\,(s+40)}. \qquad \ldots(3.42)$$

Solution. For the response of $T_1(s)$, multiply by $\frac{1}{s}$ and take partial fractions.

$$\frac{2000}{(s+20)^2\,(s+40)} = \frac{A}{(s+20)^2} + \frac{B}{(s+20)} + \frac{C}{(s+40)} \qquad \ldots(3.43)$$

A and C can be found by the. cover up rule to give $A = 100$ and $C = 5$. Then cross-multiplying expression (3.43),

$$2000 = A\,(s+40) + B\,(s+20)\,(s+40) + C\,(s+20)^2 \qquad \ldots(3.44)$$

Equating the coefficients of s^2, $0 = B + C$. Thus $B = -C = -5$. (This may be checked by equating the coefficients of s^1 and s^0.)
Thus, referring to the transform tables,

$$r_1(t) = 100te^{-20t} - 5e^{-20t} + 5e^{-40t} \qquad \ldots(3.45)$$

For $R_2(s)$,

$$\frac{2000}{(s+19)\,(s+21)\,(s+40)} = \frac{A}{(s+19)} + \frac{B}{(s+21)} + \frac{C}{(s+40)} \qquad \ldots(3.46)$$

By cover up,

$$A = 47.6, \quad B = -52.6, \quad C = 5.01,$$

and

$$r_2(t) = 47.6e^{-19t} - 52.6e^{-21t} + 5.01e^{-40t}\,. \qquad \ldots(3.47)$$

Substituting values of t between 0.01 and 0.5 into $r_1(t)$ and $r_2(t)$ will show that, to three significant figures, they are identical. A sketch of this common result is shown in Fig. 3.10.

An alternative approximation would be a transfer function with a complex conjugate pair of poles instead of the double pole. In this case, the approximate function would be:

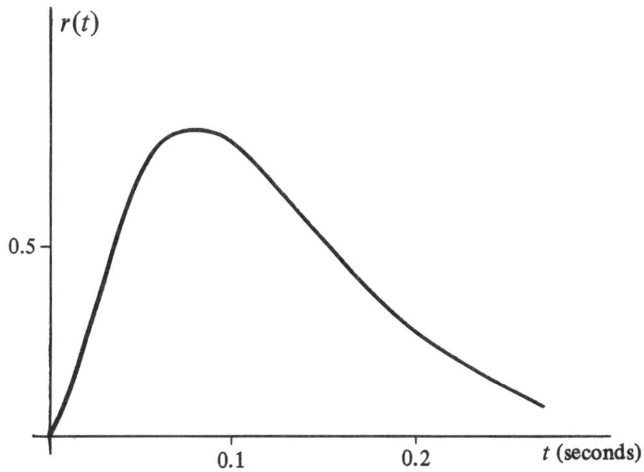

Fig. 3.10 The step response for a transfer function having a coincident pair of real negative poles.

$$R_3(s) = \frac{2000}{(s+20-j)(s+20+j)(s+40)} \quad \ldots (3.48)$$

The resulting response would be of the form:

$$r_3(t) = A e^{-20t} \sin(t+\phi) + B e^{-40t} , \quad \ldots (3.49)$$

and once again, it would be virtually the same as $r_1(t)$ and $r_2(t)$. Of the three alternatives, the approximate $r_2(t)$ is the most convenient as the residues can be found by the cover up method and the result is more easily plotted or sketched.

3.11 COMPLEX CONJUGATE POLES

We shall now consider transfer functions having a complex conjugate pair of poles. As a result, the response will include an exponentially decaying sinusoid. The first group of this type has no other poles, and a zero at the origin of the transfer function. The step response pole zero diagram will include only the complex pair of poles. A response function of this type would be obtained from the *LCR* circuit shown in Fig. 3.11a or in a second-order feedback system. In either case, the response $R(s)$ has the form:

$$R(s) = \frac{a}{s^2 + as + b} \quad \ldots (3.50)$$

In order that different types of system can be compared, it is convenient to express this response function in terms of two different parameters. These are the system natural or undamped resonant frequency ω_n and the damping factor ζ. The resulting form for $R(s)$ is then given by:

$$R(s) = \frac{2\zeta\omega_n}{s^2 + 2\zeta\omega_n s + \omega_n^2} \,. \qquad \ldots(3.51)$$

The poles of this function lie at $-\zeta\omega_n \pm \omega_n\sqrt{1-\zeta^2}$. This is shown in Fig. 3.11b.

Analysis of the circuit in Fig. 3.11a shows that the constants ω_n and ζ are given by:

$$\omega_n = \frac{1}{\sqrt{LC}} \quad \text{and} \quad \zeta = R\sqrt{\frac{C}{4L}} \,. \qquad \ldots(3.52)$$

It is useful to notice that the Q factor, which may be written:

$$Q = \frac{\omega_n L}{R} \quad \text{or} \quad Q = \frac{1}{R}\sqrt{\frac{L}{C}} \qquad \ldots(3.53)$$

may be compared with (3.52) to show that $Q = \frac{1}{2\zeta}$ . ...(3.54)

The two variables ζ and Q are introduced here since control engineers view ζ as a measure of system *damping,* whereas communications engineers or those associated with signal processing or filtering are inclined to use Q as a measure of the network's frequency selectivity.

Clearly, having obtained $R(s)$ or $T(s)$, then simply equating to (3.50) permits the Q factor, bandwidth and resonant frequency to be determined by inspection:

$$\omega_n = \sqrt{b}$$

$$Q = \sqrt{\frac{b}{a}}$$

$$BW = \frac{a}{2}$$

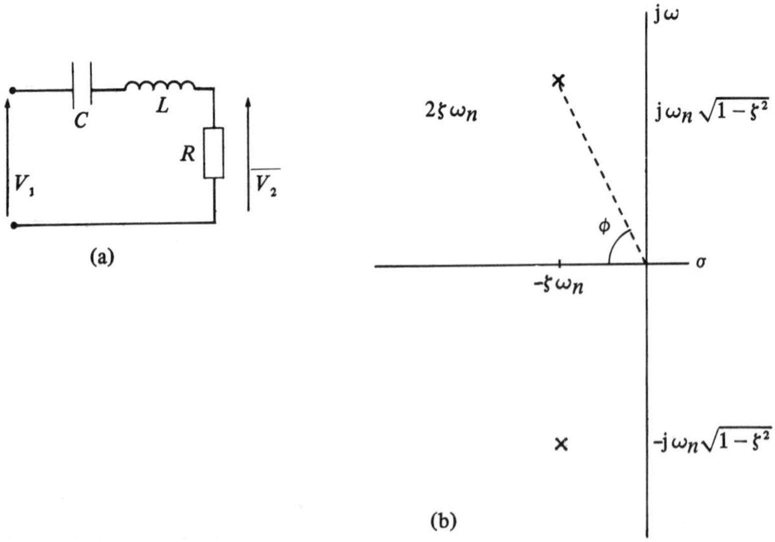

Fig. 3.11 An *LCR* circuit providing a pair of complex conjugate poles. a) The circuit. b) The pole zero diagram showing the natural resonant frequency ω_n and the damping factor ζ.

In order that comparisons may be made between the responses associated with different pole zero diagrams, it is convenient to consider response functions having the same imaginary component $\omega_n\sqrt{1-\zeta^2}$, but at different angles ϕ, with respect to the real axis. Referring to Fig. 3.11b, the length of the broken line from the origin to the pole is $\omega_n\sqrt{\zeta^2+1-\zeta^2}$ which is simply ω_n. Thus, the angle ϕ is $\cos^{-1}\dfrac{\zeta\omega_n}{\omega_n}$,

$$\therefore \qquad \zeta = \cos\phi\,, \qquad \ldots(3.55)$$

and

$$Q = \frac{1}{2\cos\phi}\,. \qquad \ldots(3.56)$$

Some functions are shown in this way in Fig. 3.12. We can see from these that the angular frequency is always given by the imaginary component of the poles and that the rate of exponential decay is given by the real component. However, if the angle ϕ is less than 60° ($\zeta < 0.5$ and $Q < 1$), the oscillatory nature is barely observable and the response then tends to the form obtained with two coincident real poles (see Example 3.3).

Another common situation is the similar transfer function without a zero at the origin. In this case a d.c. component will appear in the response. Fig. 3.13 shows an *RLC* circuit for which this applies. For this arrangement, the values of ζ and ω_n are the same as for the circuit in Fig. 3.11 but note that the numerator is now $\omega_n{}^2$.

Another system having the same form of transfer function is the simple position control system described in Example 2.12.

$$T(s) = \frac{\theta_o}{\theta_i}(s) = \frac{K/J}{s^2 + sF_v/J + K/J}\,. \qquad \ldots(3.57)$$

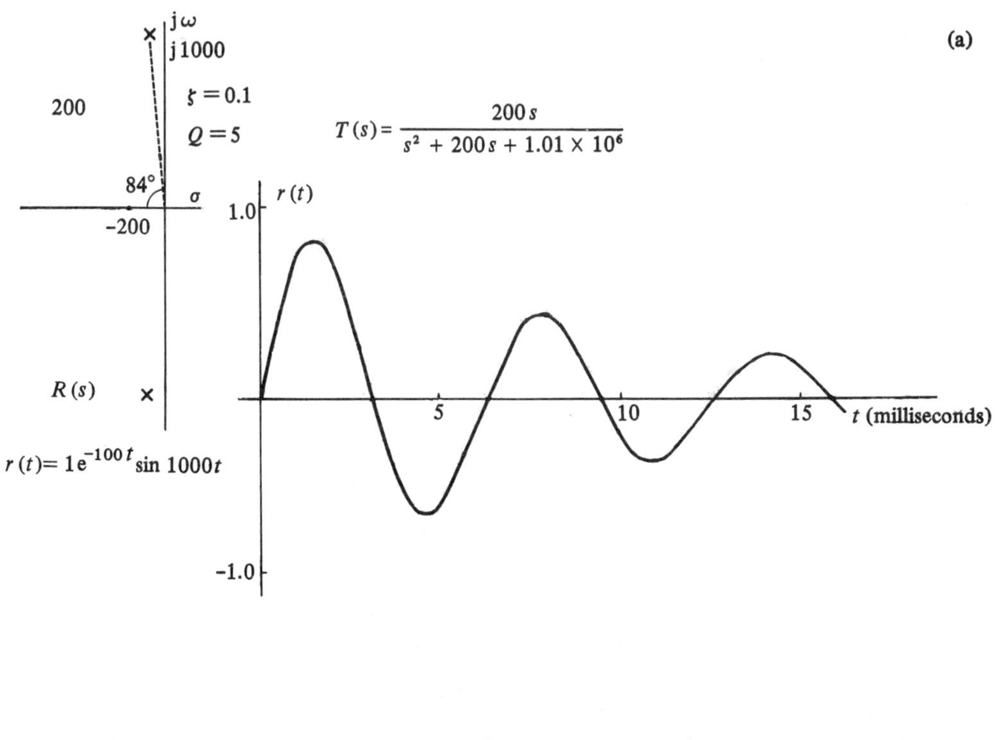

(b)

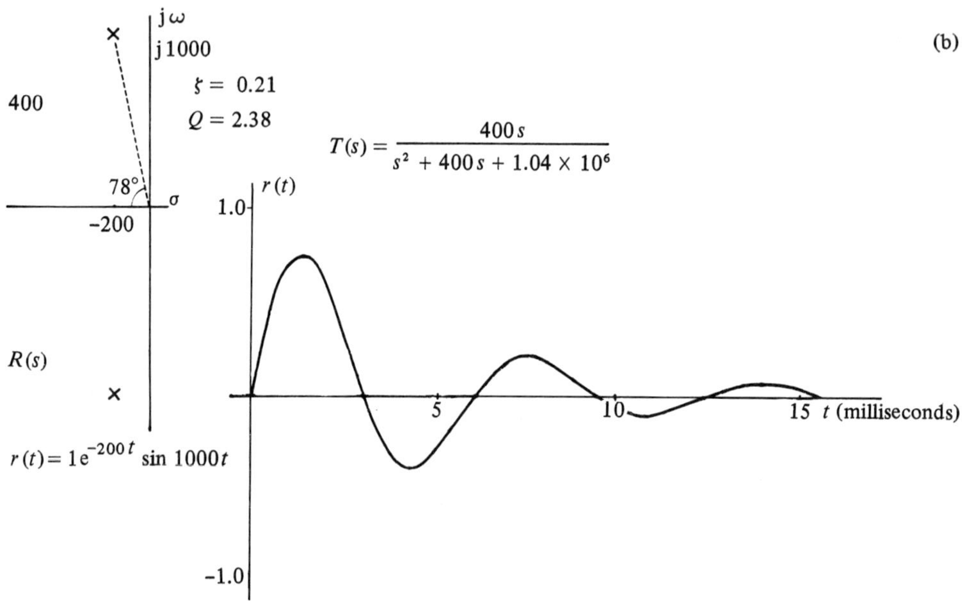
jω
j1000
ζ = 0.21
Q = 2.38
400
$T(s) = \frac{400s}{s^2 + 400s + 1.04 \times 10^6}$
78°
σ
-200
r (t)
1.0
R(s)
5
10
15
t (milliseconds)
$r(t) = 1e^{-200t} \sin 1000t$
-1.0

(c)

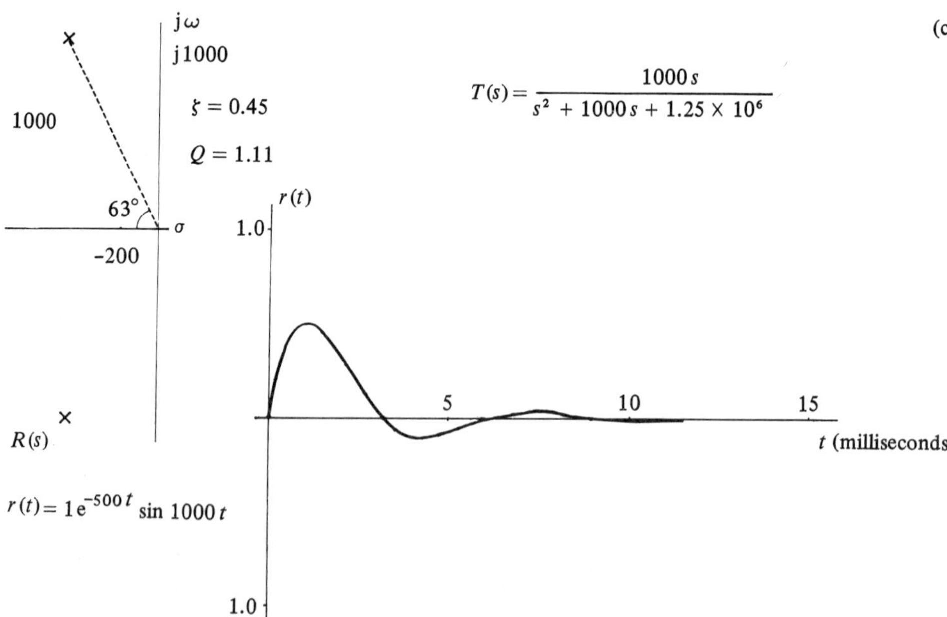
jω
j1000
ζ = 0.45
Q = 1.11
1000
$T(s) = \frac{1000s}{s^2 + 1000s + 1.25 \times 10^6}$
63°
σ
-200
r (t)
1.0
5
10
15
R(s)
t (milliseconds)
$r(t) = 1e^{-500t} \sin 1000t$
1.0

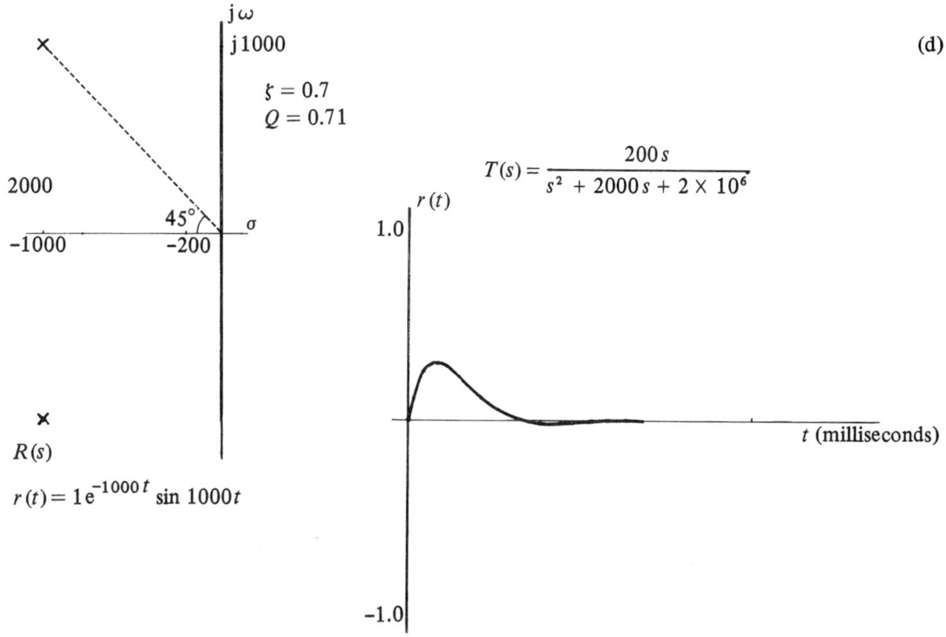

Fig. 3.12 Some response functions having only a single pair of complex conjugate poles.

The standard form for this system and for the circuit shown in Fig. 3.13 is given by:

$$T(s) = \frac{\omega_n^2}{s^2 + 2\zeta\omega_n s + \omega_n^2} \qquad \ldots (3.58)$$

For the control system, the values for the undamped resonant frequency and the damping factor are:

$$\omega_n = \sqrt{K/J} \quad \text{and} \quad \zeta = \frac{F_v}{2\sqrt{JK}}. \qquad \ldots (3.59)$$

A single example is sufficient to show how the step response for this type of transfer function differs from those shown in Fig. 3.12. This example, shown in Fig. 3.14, shows the presence of a d.c. level as well as the oscillatory component. There are some other useful points to note about this response.

i. The angle of the sine term is such that at $t = 0$, the response $r_{(0)} = 0$ ($1.02e^0 \sin 102° = 1$).

ii. The time T_1 to the peak of the first overshoot is exactly half the period of oscillation $\frac{2\pi}{\omega}$, i.e. $T_1 = \frac{\pi}{\omega}$ where $\omega = \omega_n\sqrt{1 - \zeta^2}$.

iii. The time T_p between positive peaks is also $\frac{2\pi}{\omega}$ regardless of the rate of decay. These results are true for all damping factors (or values of Q) which may be proved by differentiation of the general form for $r(t)$.

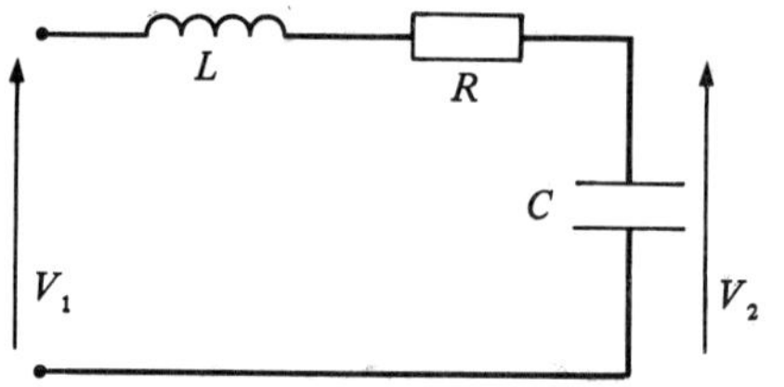

$$T(s) = \frac{\omega_n^2}{s^2 + 2\zeta\omega_n s + \omega_n^2}$$

$$\omega_n = \frac{1}{\sqrt{LC}},\quad \zeta = \frac{R}{2}\sqrt{\frac{C}{L}},\quad Q = \frac{1}{R}\sqrt{\frac{L}{C}}$$

Fig. 3.13 A series resonant circuit having only a pair of complex conjugate poles.

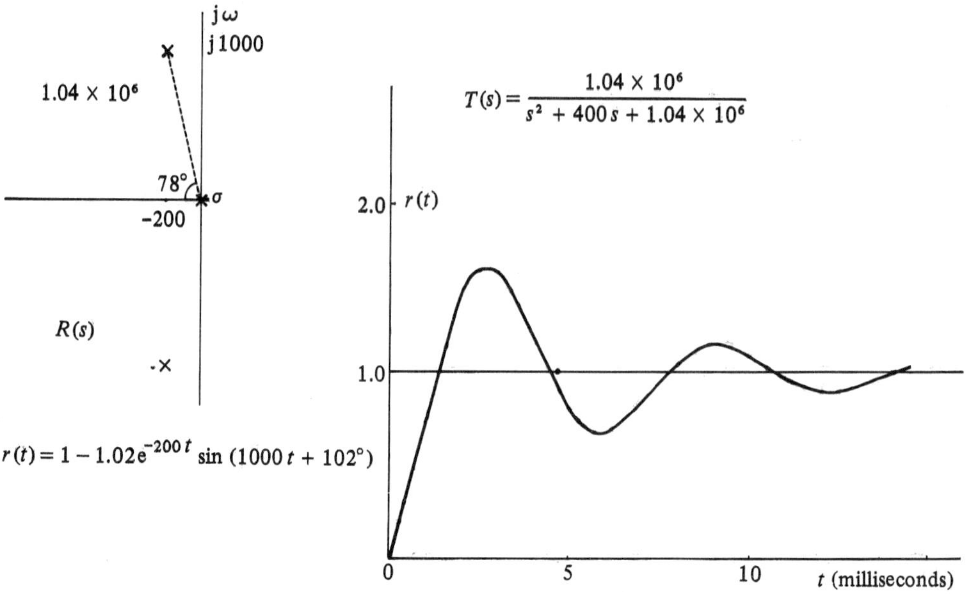

Fig. 3.14 The step response and pole zero diagram for a system having only a pair of complex conjugate poles.

3.12 COMPLEX POLES, REAL POLES AND ZEROS

The final group of transfer functions that we need consider in this section are those that have a complex conjugate pair of poles and additional real poles and zeros. Some examples of response pole zero diagrams, transfer functions and response waveforms are shown in Fig. 3.15. Circuits or systems have not been shown in these cases but they could be considered as combinations of previous examples where interaction has been avoided by the use of isolating amplifiers. In practice, they are more likely to occur in feedback systems.

(a)

$$T(s) = \frac{5.05 \times 10^7}{(s+50)(s^2+200s+1.01\times10^6)}$$

$$r(t) = 1 - 1.0075e^{-50t} - 0.05e^{-100t}\sin(1000t - 9°)$$

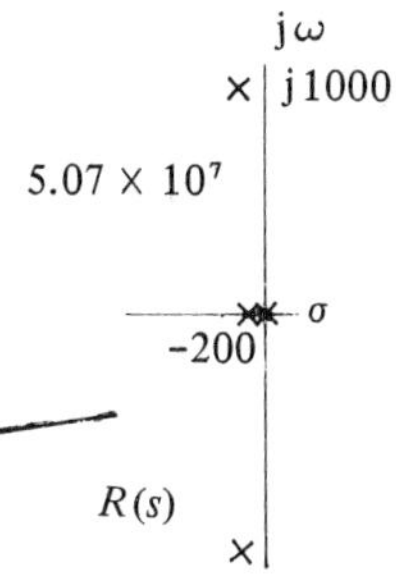

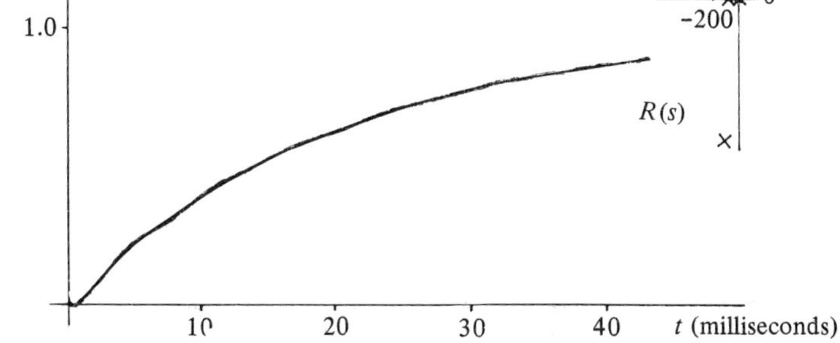

(b)

$$T(s) = \frac{2.53 \times 10^8}{(s+250)(s^2+200s+1.01\times10^6)}$$

$$r(t) = 1 - 0.988e^{-250t} + 0.248e^{-100t}\sin(1000t - 3°)$$

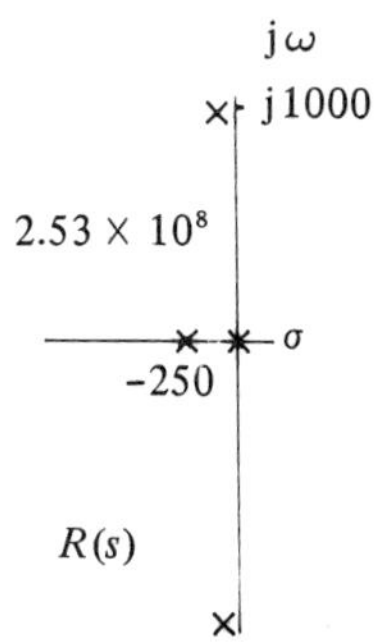

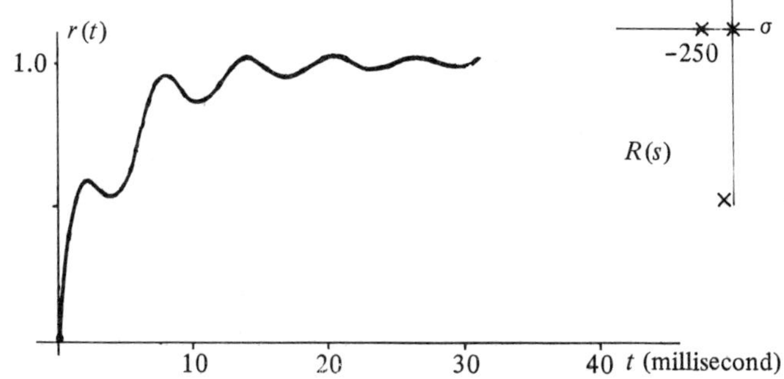

(c)

$$T(s) = \frac{6.31 \times 10^5 (s+80)}{(s+50)(s^2+200s+1.01\times10^6)}$$

$$r(t) = 1 - 0.38e^{-50t} + 0.626e^{-100t}\sin(1000t - 82.5°)$$

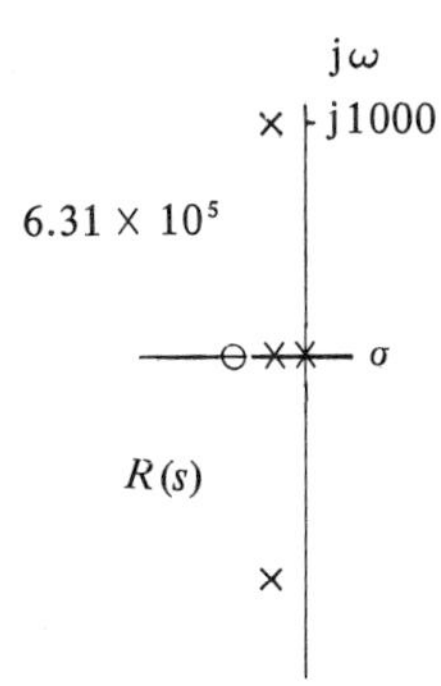

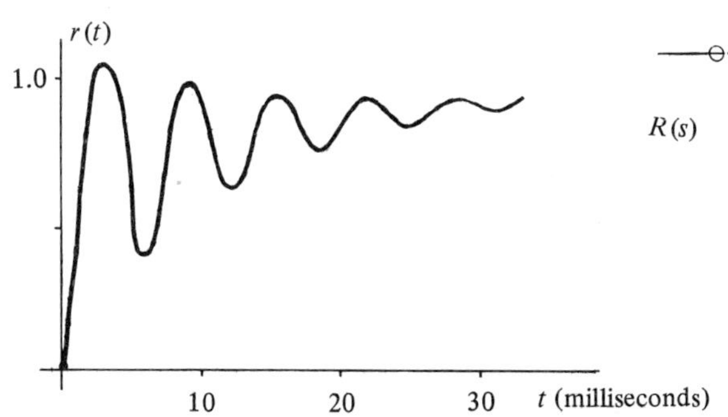

Fig. 3.15 Some step response functions having complex poles and real poles and zeros.

In case (a), the step and the real system pole produce the familiar exponent towards unity. The complex conjugate poles add a decaying sinusoid to this; the amplitude, however, is small, as these poles are far from the origin as compared with the real pole. The small phase angle results in the usual total response of zero at $t = 0$.

In case (b), the real pole is further from the origin and the time constant of the resulting exponent is shorter. In addition, the relative amplitude of the decaying sinusoid is increased.

With (c), the real pole is in the same place as in case (a), but its effect is reduced by the addition of a real zero. This has reduced the amplitude of the exponent and at the same time it has increased the amplitude of the decaying sinusoid. Once again, at $t = 0$, the combined effect is to make $r(t) = 0$.

3.13 THE TRANSIENT RESPONSE TO OTHER SWITCHED INPUTS

Although the transient response to a step input is important, other input signals may also be 'switched on' at $t = 0$. Important examples of these are switched sinusoids and switched ramps. Another transient situation is the case of a very short pulse which can be approximated to an impulse. All these cases can be represented on the response pole zero diagram and in most situations, the response $r(t)$ can be obtained using only the methods discussed above. This will now be illustrated by some examples.

Example 3.4. Determine the response of the network shown in Fig. 3.8b to the following switched inputs:

$$\text{a)}\quad v_i = 10 \sin 1000t,$$

$$\text{b)}\quad v_i = 10 \cos 1000t,$$

$$\text{c)}\quad v_i = 1000t.$$

Solution, with reference to Table 1.1. The transforms for the three required inputs are respectively $\dfrac{10^4}{s^2 + 10^6}$, $\dfrac{10s}{s^2 + 10^6}$, and $\dfrac{1000}{s^2}$. The required network voltage transfer function is given by: $T(s) = \dfrac{s + 1000}{s + 6000}$.

After combining the information, the three response pole zero diagrams can be drawn as shown in Fig. 3.16. Note that in each case, the system pole is still present and we can therefore expect the same transient exponent to appear in the response. In addition, there will be terms due to the applied signal. In the first two cases, these can be analysed in the same way as before, after obtaining the partial fraction residues by means of the directed line segment method.

$$R(s) = \frac{10^4(s + 1000)}{(s^2 + 10^6)(s + 6000)} = \frac{A}{s + 6000} + \frac{B}{s - \text{j}1000} + \frac{B^*}{s + \text{j}1000} \quad \ldots (3.60)$$

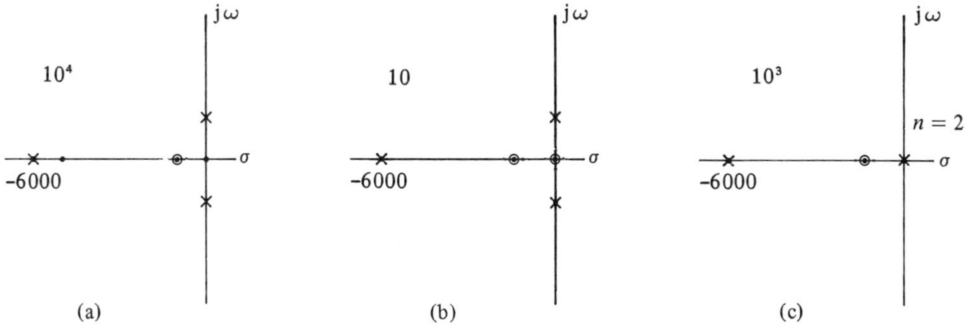

Fig. 3.16 The response pole zero diagram for Example 3.4.

From the pole zero diagram,

$$A = \frac{10^4 \times (-5000)}{1000^2 + 6000^2} = -1.35 \; ; \qquad \ldots (3.61)$$

$$B = \frac{10^4 \times 1000 \sqrt{2}\angle 45^\circ}{\sqrt{6000^2 + 1000^2} \angle \tan^{-1}\left(\frac{1}{6}\right) \times 2000 \angle 90^\circ} = \frac{1.16\angle 35.5^\circ}{\angle 90^\circ} \; . \qquad \ldots (3.62)$$

Taking the inverse transform,

$$r(t) = -1.35e^{-6000\,t} + 2.32 \sin (1000t + 35.5^\circ) \; . \qquad \ldots (3.63)$$

Notice that in this example, the fact that the complex terms have no real parts does not invalidate the methods used before. The solution indicates both a transient and steady-state term. In addition, it may be noted that the amplitude of the steady-state sinusoidal term derives from the excitation frequency, and will change according to the signal frequency. This leads to the notion of frequency selective networks.

The method for this input is identical to the last case, so the solution is left for the reader. The resulting response is given by:

$$r(t) = 8.1e^{-6000\,t} + 2.32 \cos (1000t + 35.5^\circ) \; . \qquad \ldots (3.64)$$

$$R(s) = \frac{1000\,(s + 1000)}{s^2 (s + 6000)} = \frac{A}{s^2} + \frac{B}{s} + \frac{C}{s + 6000} \; . \qquad \ldots (3.65)$$

In this case, the double pole at the origin requires the extra term in the partial fraction expansion. Residues A and C can be found from the pole zero diagram but, we must work from first principles to determine B. From the pole zero diagram,

$$A = \frac{1000 \times 1000}{6000} = 167 \qquad \ldots (3.66)$$

$$C = \frac{1000 \times (-5000)}{6000^2} = -0.139 \qquad \ldots (3.67)$$

Multiplying equation (3.65) by $s^2(s + 6000)$,

$$1000\,(s + 1000) = A\,(s + 6000) + Bs\,(s + 6000) + Cs^2 \quad \ldots (3.68)$$

Equating the coefficients of s^2,

$$B = -C = 0.139.$$

$$\therefore \quad r(t) = 167t + 0.139 - 0.139e^{-6000\,t} \quad \ldots (3.69)$$

Comparing the results for the three signals, we can see that the term $e^{-6000\,t}$ appears in each case. Such components, resulting from the system poles, will always be present regardless of the input signal. The amplitude and sign of these terms will be determined, in part, by the nature of the applied signal. The response may also include terms related to the applied signal. Where this signal is a sinusoid, the general method described earlier in the chapter can be applied but with a ramp input, the resulting double pole necessitates first principles partial fraction expansion unless the transfer function includes one or more zeros at the origin.

3.14 THE IMPULSE RESPONSE

Systems will sometimes be subject to excitation by very short pulses. If the duration of a pulse is much less than the system time constants, the shape of the pulse is unimportant. The response then depends upon the energy contained in the pulse. An impulse is an exciting function which fits this description and one for which the Laplace transform is unity. Thus, if an impulse is applied to a system having a transfer function $T(s)$, the response $R(s)$ is simply $T(s) \times 1$ and can be evaluated in the normal manner. It will of course, contain only terms due to the poles of $T(s)$.

3.15 SUMMARY

In conclusion, we will summarize the various effects of poles and zeros upon the transient response of circuits and systems.

i. Every pole in the response function will result in a component in the time response $r(t)$.
ii. A pole at the origin produces a d.c. step and thus a d.c. level after all transient changes are completed.
iii. Real negative poles produce decaying exponents, of the form $Ae^{-\sigma t}$, whose time constants are the reciprocal of the pole frequency σ.
iv. A double pole at the origin results in a ramp with a slope given by the residue at the pole, together with a d.c. shift.
v. Complex conjugate pairs of poles with negative real parts result in exponentially decay-

ing or damped sinusoids whose angular frequencies are given by the imaginary components $j\omega$: the rates of decay of these sinusoids have time constants corresponding to the reciprocals of the real parts of the pole frequencies.

vi. A pair of poles at $\pm j\omega$ (no real parts), represents a steady oscillation at the angular frequency ω.

vii. Complex conjugate poles with positive real parts or real positive poles result in response terms that increase exponentially towards infinity. Any system having such poles is therefore said to be unstable.

3.16 AMPLITUDE OF RESPONSE COMPONENTS OR RESIDUES

i. The amplitude and sign or phase angle of each response term described above, depends upon the position of the pole in question, relative to all the other poles and zeros of the function.

ii. Poles which are near each other have residues which are large compared with those of other poles which are further away.

iii. A zero which is near a particular pole will reduce the residue of that pole and increase the residues of other poles. Taken to the limit, a pole and zero in the same place, cancel each other.

iv. The sign of a residue for a real pole depends upon the number of poles and zeros to the right of the pole in question. An odd number to the right makes the residue negative and an even number (including 0) results in a positive component. Complex conjugate pairs of poles or zeros have no effect on this result as the angles from such pairs always cancel.

v. The residues may be found (or estimated) from the general rule:

$$A = K \frac{\left[\begin{array}{l}\text{The product of the directed line segments from each zero}\\ \text{to the pole in question}\end{array}\right]}{\left[\begin{array}{l}\text{The product of the directed line segments from each other}\\ \text{pole to the pole in question.}\end{array}\right]}$$

where A is the residue at the pole in question.

vi. For a complex conjugate pair of poles, if the residue B is found for the pole at $(-\sigma + j\omega)$, the response term for the pair is given by:

$$r(t) = |2jB|e^{-\sigma t} \sin(\omega t + \angle 2jB)$$

3.17 SPECIAL RESULTS FOR A STEP INPUT

i. Transfer functions having no pole or zero at the origin.

a) If $T(s)$ has no zeros, $r(t)$ will start from zero, i.e. there will be no initial step. It will also finish at a d.c. level after all exponentials have decayed.

b) If $T(s)$ has as many zeros as there are poles, there will be an initial step in $r(t)$; if there are less zeros than poles, there will be no step. With real circuits and systems, it is possible for the transfer function to have one zero more than the number of poles: in such cases, the step response function will have an equal number of poles and zeros and in general, there will be a step at $t = 0$.

c) Results (a) and (b) above still apply when the poles include a complex conjugate pair.

ii. Transfer functions having a zero at the origin.
The step response will have a final d.c. level of zero, but otherwise the rules outlined above are followed. The zero at the origin also represents differentiation and can often usefully be interpreted in this way. In particular, a ramp input will produce a d.c. level in the output.

iii. Transfer functions having a pole at the origin.
With a step input, there will be a ramp in the output. This will, of course, tend to infinity as t tends to infinity, however small the input step is. If the input is zero, the output will settle to a steady d.c. level. The pole at the origin also represents integration and is often found in the transfer functions of systems used for analogue computing, simulation, signal processing and control.

iv. In all cases, the value of $r(t)$ as $t = 0$, may be found by letting s tend to infinity in $R(s)$, the expression for the response.

v. In all cases, the final value of $r(t)$, as t tends to infinity, may be obtained by substituting $s = 0$ into $R(s)$.

Note. The response function, $R(s)$, for a unit step type of excitation is related to the transfer function, $T(s)$, by:

$$R(s) = \frac{1}{s} T(s) ,$$

hence,

$$T(s) = sR(s) .$$

Now the infinitely fast transition at $t = 0$, associated with the step input, implies the network is excited by high frequencies and therefore the behaviour of $T(s)$ as $s \to j\infty$ is indicative of how the network or system reacts to the excitation at $t = 0$. Likewise, the d.c. or constant nature of the step input as $t \to \infty$ implies the network is excited by low frequencies or d.c. and therefore the behaviour of $T(s)$ as $s \to 0$ is an indication as to how the system reacts to the d.c. type of excitation as $t \to \infty$. These two properties can be summarized mathematically by the initial and final value theorems:

$$\lim_{s \to \infty} T(s) = \lim_{s \to \infty} sR(s) = r(0^+)$$

and

$$\lim_{s \to 0} T(s) = \lim_{s \to 0} sR(s) = \lim_{t \to \infty} r(t)$$

respectively.

Thus, by simply examining $T(s)$, or equivalently $sR(s)$, and putting $s \to 0$ or $s \to \infty$ allows the initial and final values of $r(t)$ to be obtained by inspection; hence serving as an effective check.

3.18 EXAMPLES FOR FURTHER PRACTICE

Example 3.5. Draw the pole zero diagram for the following transfer functions.

a) $$\frac{(s+8)(s+20)}{(s+2)(s+80)}.$$

b) $$\frac{20(s+3)}{s(s^2+4s+13)}.$$

c) $$\frac{150(s-200)}{s^2(s+500)}.$$

d) $$\frac{10^6(s^2+6000s+5\times 10^6)}{(s^2+10^4 s+4\times 10^8)}.$$

e) $$\frac{(s+10)(s+20)}{(s^2+100)}.$$

Example 3.6. Draw the response pole zero diagram when a step input is applied to the systems having the following transfer functions.

a) $$\frac{8\times 10^9}{(s+10^4)(s+10^5)}.$$

b) $$\frac{(s+8)(s+20)}{(s+2)(s+80)}.$$

Hence determine the time domain step response.

$$\left(8 - 8.9e^{-10^4 t} + 0.9e^{-10^5 t},\ 1 - \frac{9}{13}e^{-2t} + \frac{9}{13}e^{-80t}\ .\right)$$

Example 3.7. Determine with the aid of pole zero diagrams, the step responses for systems having the following transfer functions.

a) $$\frac{10}{s^2+2s+10}.$$

b) $$\frac{13.2\times 10^5(s+100)}{(s+500)(s^2+120s+66\,100)}.$$

c) $$\frac{2.5s\,(s^2+2s+26)}{(s^2+8s+25)}\,.$$

$(1 + 1.054e^{-t} \sin(3t - 108.4°); 4 + 4.13e^{-500t} + 10.29e^{-60t} \sin(250t - 52.2°)$;

$25.67e^{-4t} \sin(3t - 36°)$.)

Example 3.8. An electronic system has a transfer function given by

$$T(s) = \frac{10^7 s}{(s+5000)^2\,(s+1000)}\,.$$

If a 10 V step is applied, determine the approximate response and hence the output voltage at 1 ms. (2.1 V.)

Example 3.9. A phase shift circuit has a transfer function given by

$$T(s) = \frac{4.8 \times 10^6}{(s+200)\,(s+2400)}\,.$$

A sinusoidal voltage of amplitude 80 V peak and frequency 50 Hz is switched to the circuit input. Determine an expression for the output voltage if the switch is closed (a) when the instantaneous voltage is zero going positive and (b) when the instantaneous voltage is at the peak negative value. Find also the amplitude and sign of the first voltage peak and the time at which it occurs.

($39.5e^{-200t} - 0.935e^{-2400t} + 42.6 \sin(314t - 65°)$, + 49 volts at 8.6 milliseconds.

$25.2e^{-200t} - 7.15e^{-2400t} + 42.6 \sin(314t + 205°)$, –30.3 volts at 3.6 milliseconds.)

Example 3.10. A ramp voltage increasing at 2×10^4 Vs^{-1} is applied to a system having a transfer function given by

$$T(s) = \frac{1.2 \times 10^3\,(s+6000)}{(s+4000)\,(s+20\,000)}\,.$$

Determine the output voltage and illustrate your answer with a sketch of the waveform.

($1.8t - 0.24 + 0.19e^{-4t} + 0.05e^{-20t}$, t in milliseconds.)

Example 3.11. Without calculation, state the important features of the step response for systems having the following transfer functions:

a) $$\frac{2 \times 10^6}{(s+400)\,(s+5000)}\,.$$

b) $$\frac{s}{(s+10^5)}\,.$$

c) $$\frac{0.4\,(s+50)}{(s+20)}\,.$$

d) $$\frac{s^2}{(s+10)\,(s+1000)}\,.$$

e) $$\frac{(s+200)\,(s+1000)}{(s+100)\,(s+2000)}\,.$$

f) $$\frac{5000s}{(s+1340)\,(s+18\,660)}\,.$$

g) $$\frac{60s}{(s^2+60s+250\,900)}\,.$$

h) $$\frac{8\times10^8}{(s+400)\,(s^2+1000s+2\times10^6)}\,.$$

i) $$\frac{1.33\times10^6(s+600)}{(s+400)\,(s^2+1000s+2\times10^6)}\,.$$

j) $$\frac{10}{(s^2+2s+10)}\,.$$

((a) Final value 1, time constants 2.5 ms negative, 0.2 ms positive, delayed exponential to final value; (b) final value zero, initial step decaying with 10μ s time constant; (c) initial step followed by exponential increase to 1 with time constant of 0.05 s; (d) initial step of 1 decaying through zero with a 1 ms time constant to a negative exponential decay to zero with a time constant of 100 ms; (e) initial step of 1 decaying with a time constant of 0.5 ms to positive exponential with time constant of 10 ms rising to 1; (f) rising from zero with an initial time constant of 54μ s then decaying to zero with a time constant of 0.75 ms; (g) sinusoid with an initial amplitude of 1, frequency 80 Hz, decaying with a time constant of 33 ms; (h) rising exponential from 0 to 1 with a time constant of 2.5 ms plus a sinusoid of frequency 210 Hz decaying with a time constant of 2 ms; (i) similar to previous case but amplitude of sinusoid is increased; (j) rising from zero to 1 with a decaying sinusoid of frequency 0.47 Hz and time constant 1 s.)

4

The Pole Zero Diagram and the Steady-State Response

4.1 INTRODUCTION

In Chapter 3, the relationship between the pole zero diagram and the transient response to switched inputs was discussed. It was shown that pole zero diagrams could be 'read' and that the general nature of a transient response could be predicted without calculation. In this chapter, we see how a similar understanding can be developed when the applied signals are d.c. or, more importantly, steady-state sinusoids. The use of transfer functions as 'gains' to specific signals is discussed; this leads to the calculation of numerical results for chosen steady-state inputs. The relationships found are compared with the pole zero diagram for the transfer function and this leads to the development of general relationships in terms of poles and zeros. As a range of signal frequencies is often considered, the frequency response is introduced in the form of the Bode plot. This too can be linked to the pole zero diagram and the effects of real and complex poles and zeros on the Bode plot are developed. Finally, a general survey of pole zero patterns with the associated frequency response is used to provide a set of guidelines for reading the frequency response from the pole zero diagram.

4.2 TRANSFER FUNCTIONS AND THE STEADY-STATE RESPONSE

In Chapter 2, the transfer function was defined as the ratio of the output quantity to the input quantity expressed as a function of s, with no initial conditions.
Thus,

$$T(s) = \frac{\text{output}}{\text{input}}(s) \ . \qquad \ldots(4.1)$$

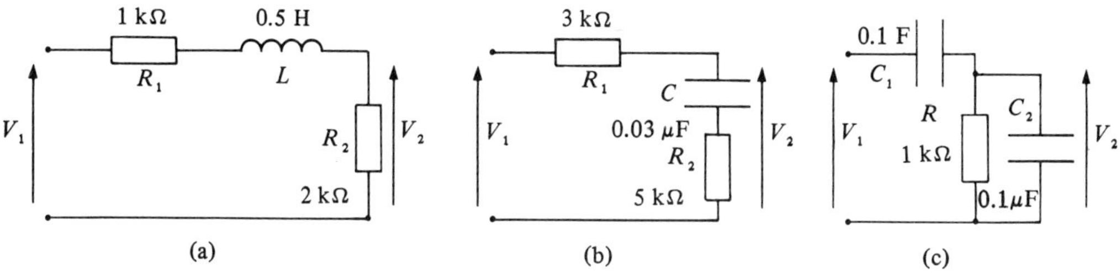

Fig. 4.1 Electrical networks for Example 4.1.

In general terms, this has the form of gain or amplification at any value of complex frequency s; where, in general $s = \sigma + j\omega$. This includes two special cases where either $s = 0$, or $\sigma = 0$ and $s = j\omega$. The first of these, ($s = 0$), was shown to represent d.c. (see Fig. 1.6, p. 8). Thus if the value $s = 0$ is substituted into $T(s)$, the result $T(0)$ is the d.c. gain of the system. This can also be seen from the final value theorem;

$$f = \lim_{t \to \infty} f(t) = \lim_{s = 0} sF(s) \qquad \ldots (4.2)$$

A step input which transforms into $\frac{A}{s}$ is simply switched d.c.: the denominator s cancels with the s in (4.2), leaving F(s) with 0 substituted for s.

Similarly, $s = j\omega$ represents a steady-state sinusoid at an angular frequency ω rads^{-1}. The result of the substitution of $j\omega$ for s in $T(s)$ provides $T(j\omega)$, the steady-state a.c. transfer function. As we shall see, this result is identical to that obtained by the use of conventional a.c. theory. In practice, it is often more convenient to calculate $T(s)$ for a circuit and, only then, to find the a.c. response, the d.c. response or the transient response as required.

Example 4.1. Determine the voltage transfer functions for the three electrical networks shown in Fig. 4.1. Hence, in each case, find the d.c. voltage gain and the a.c. voltage gain to an input with an angular frequency of 5000 rads^{-1}.

Solution. In each case, the transfer function can be found by simple series parallel relationships and the results, which should be verified by the reader, are:

$$T_a(s) = \frac{R_2/L}{s + \dfrac{(R_1 + R_2)}{L}}; \qquad \ldots (4.3)$$

$$T_b(s) = \frac{R_2}{R_1 + R_2} \; \frac{(s + 1/CR_2)}{(s + 1/C(R_1 + R_2))}; \qquad \ldots (4.4)$$

$$T_c(s) = \frac{sC_1/(C_1 + C_2)}{s + 1/R(C_1 + C_2)}. \qquad \ldots (4.5)$$

For the d.c. voltage gain, we allow s to become zero and find:

$$T_a(0) = \frac{R_2}{R_1 + R_2} ,$$

and substituting values,

$$T_a(0) = 2/3 , \qquad \ldots(4.6)$$

$$T_b(0) = 1 ; \qquad \ldots(4.7)$$

$$T_c(0) = 0 . \qquad \ldots(4.8)$$

To obtain the a.c. gain at $\omega = 5000$, we substitute j5000 for s in each case and rearrange the expression into the Bode form.

$$T_a(\mathrm{j}\omega) = \frac{R_2}{R_1 + R_2}\left(\frac{1}{1 + \dfrac{\mathrm{j}\omega L}{R_1 + R_2}}\right); \qquad \ldots(4.9)$$

$$T_a(\mathrm{j}5000) = \frac{2}{3}\left(\frac{1}{1 + \mathrm{j}0.83}\right) = 0.86\angle{-40^\circ} ; \qquad \ldots(4.10)$$

$$T_b(\mathrm{j}\omega) = \frac{1 + \mathrm{j}\omega CR_2}{1 + \mathrm{j}\omega C(R_1 + R_2)} ; \qquad \ldots(4.11)$$

$$T_b(\mathrm{j}5000) = \frac{1 + 10.75}{1 + \mathrm{j}1.2} = 0.64\angle{-13^\circ} ; \qquad \ldots(4.12)$$

$$T_c(\mathrm{j}\omega) = \frac{\mathrm{j}\omega C_1 R}{1 + \mathrm{j}\omega R(C_1 + C_2)} ; \qquad \ldots(4.13)$$

$$T_c(\mathrm{j}5000) = \frac{\mathrm{j}0.5}{1 + \mathrm{j}} = 0.35\angle 45^\circ . \qquad \ldots(4.14)$$

The algebraic manipulations have not been included in this example but it would provide useful practice if the reader ensures that the rearrangements from (4.3) to (4.9), (4.4) to (4.11) and (4.5) to (4.13) can be accomplished.

4.3 THE GENERAL FORM OF $T(s)$ AND THE STEADY-STATE RESPONSE

One of the general forms of $T(s)$ has been given in Chapter 2, expression (2.4).

$$T(s) = K\frac{(s + z_1)(s + z_2)\cdots(s + z_n)}{(s + p_1)(s + p_2)\cdots(s + p_m)} . \qquad \ldots(4.15)$$

From this expression, we can find the d.c. gain $T(0)$ and the a.c. gain $T(\mathrm{j}\omega)$.

$$T(0) = K\frac{z_1 z_2 \cdots z_n}{p_1 p_2 \cdots p_m} \quad \ldots (4.16)$$

Note that in some cases, either z_1 may be zero resulting in a zero d.c. gain or p_1 may be zero, in which case the d.c. gain is infinite (i.e. integration).

$$T(j\omega) = K\frac{(j\omega + z_1)(j\omega + z_2)\cdots(j\omega + z_n)}{(j\omega + p_1)(j\omega + p_2)\cdots(j\omega + p_m)} \quad \ldots (4.17)$$

Each of the product terms in this expression is a phasor having modulus and angle (or phase). For example, if $z_1 = -a + jb$, then $j\omega + z_1 = -a + j(\omega + b)$. Then

$$|j\omega + z_1| = \sqrt{a^2 + (\omega + b)^2}, \quad \ldots (4.18)$$

and

$$\angle j\omega + z_1 = \tan^{-1}\frac{\omega + b}{-a}. \quad \ldots (4.19)$$

In many problems and practical applications, only the modulus of the gain is required. This is obtained from result (4.17) in the form:

$$T(j\omega) = K\frac{|j\omega + z_1|\cdot|j\omega + z_2|\cdots|j\omega + z_n|}{|j\omega + p_1|\cdot|j\omega + p_2|\cdots|j\omega + p_m|}. \quad \ldots (4.20)$$

Occasionally, the angle is required and it can be obtained from:

$$\angle T(j\omega) = [\angle j\omega + z_1 + \angle j\omega + z_2 + \cdots \angle j\omega + z_n] - [\angle j\omega + p_1 + \angle j\omega + p_2 \cdots \angle j\omega + p_m]. \quad \ldots (4.21)$$

Results (4.20) and (4.21) can be evaluated directly for specified values of ω as is illustrated in the next example. However, it is usually more convenient to express such information as a frequency response. This is considered in detail later in the chapter.

Example 4.2. The transfer function for the voltage gain of a feedback amplifier is given by:

$$T(s) = \frac{10^6 s(s + 1000)}{(s + 10)(s^2 + 1000s + 1.0025 \times 10^8)}. \quad \ldots (4.22)$$

Determine the modulus and angle of the gain at a) d.c., b) $\omega = 20$ and $\omega = 10^4$.

Solution. a) for the d.c. gain, we note that the first zero is at $s = 0$ resulting in a d.c. gain of 0.

b) The roots of the quadratic are $-500 \pm j10^4$ so $T(s)$ can be rewritten:

$$T(s) = \frac{10^6 s(s + 1000)}{(s + 10)(s + 500 - j10^4)(s + 500 + j10^4)}. \quad \ldots (4.23)$$

Substituting for $s = \mathrm{j}20$,

$$T(\mathrm{j}20) = \frac{10^6 \times \mathrm{j}20 \times (1000 + \mathrm{j}20)}{(10 + \mathrm{j}20)\,(500 - \mathrm{j}9980)\,(500 + \mathrm{j}10\,020)} \qquad \ldots (4.24)$$

$$= 8.92\angle 117^\circ$$

c) The substitution of $s = \mathrm{j}10^4$ into expression (4.23) is left for the reader, but the result is:

$$T(\mathrm{j}10^4) = 1005\angle{-4.2^\circ} \;. \qquad \ldots (4.25)$$

4.4 THE POLE ZERO DIAGRAM AND THE STEADY-STATE RESPONSE

In Chapter 3, the pole zero diagram was shown to be a 'map' of a transfer function or of a response function. The pole zero diagram can also be used as a map to read the response to a steady-state sinusoidal input.

The transfer function is an expression for the gain of a circuit or system at any complex frequency $s = \sigma + \mathrm{j}\omega$. For the steady-state sinusoidal response, the required value of s is simply $\mathrm{j}\omega$. This value of s can be plotted on the pole zero diagram for any chosen value of angular frequency ω.

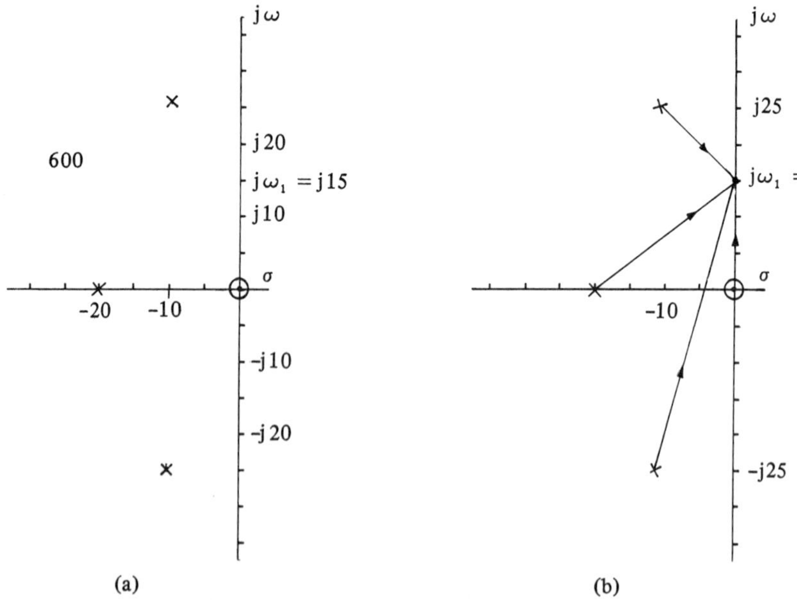

Fig. 4.2 The pole zero diagram and the steady-state response. a) The poles and zeros of $T(s)$. b) The directed line segments for $T(\mathrm{j}15)$.

Consider the transfer function:

$$T(s) = \frac{600s}{(s+20)(s+10-\mathrm{j}25)(s+10+\mathrm{j}25)} \;. \qquad \ldots (4.26)$$

The pole zero diagram is shown in Fig. 4.2a and in addition, a value of $\mathrm{j}\omega = \mathrm{j}15$ is indicated. Applying (4.17) above,

$$T(\mathrm{j}15) = \frac{600\,(\mathrm{j}15)}{(20+\mathrm{j}15)(10-\mathrm{j}10)(10+\mathrm{j}40)} \;. \qquad \ldots (4.27)$$

In Fig. 4.2b, the directed line segments have been drawn *from* the three poles and *from* the single zero to the point j15. Examination of the four phasors in expression (4.27) shows that each phasor corresponds to the directed line segment from the pole or zero which produced that term in $T(\mathrm{j}15)$. For example, the term $(s + 20)$ resulted in the pole at $s = -20$. This, in turn, produced the term $(20 + \mathrm{j}15)$ in $T(\mathrm{j}15)$. Finally, the directed line segment from the pole to j15 is, by inspection, $(+20 + \mathrm{j}15)$. In a similar way, the remaining directed line segments correspond to the other terms in $T(\mathrm{j}15)$. In many instances, only $|T(\mathrm{j}\omega)|$ is required and this can be obtained generally, in terms of the directed line segments as follows:

For a given transfer function represented by a pole zero diagram, the modulus of the steady-state sinusoidal gain $|T(\mathrm{j}\omega)|$ is given by:

$$|T(\mathrm{j}\omega)| = K\,\frac{\begin{bmatrix}\text{The product of the directed line segments}\\ \textit{from}\text{ each zero }\textit{to}\text{ the specified j}\omega.\end{bmatrix}}{\begin{bmatrix}\text{The product of the directed line segments}\\ \textit{from}\text{ each pole }\textit{to}\text{ the specified j}\omega.\end{bmatrix}} \;. \qquad \ldots (4.28)$$

Should the angle be required, this is obtained from the angles of the directed line segments *from* the poles and zeros *to* jω.

$\angle T(\mathrm{j}\omega)$ = [Sum of the angles *from* the zeros *to* jω] – [Sum of the angles *from* the poles *to* jω] . ...(4.29)

Results (4.28) and (4.29) can both be applied to the more general case of $T(s)$ to find $|T(s)|$ and $\angle T(s)$ for any value of complex frequency s. These results are used in the next chapter for root locus plotting.

It is useful to compare results (4.28) and (4.29) with the similar result (3.12) on page 74. In (4.28), we require directed line segments from *all* the poles and zeros; in (3.12), the directed line segments are from all the zeros and from each *other* pole to the pole in question. We can now examine the application of result (4.28) by reference to some examples.

Example 4.3. Fig. 4.3 shows an *LCR* circuit with normalized component values (see Appendix A3). Determine the voltage transfer function for the circuit and with reference to a pole zero diagram investigate the frequency response. Find also the phase shift introduced by the circuit at a frequency of 1 rads^{-1}.

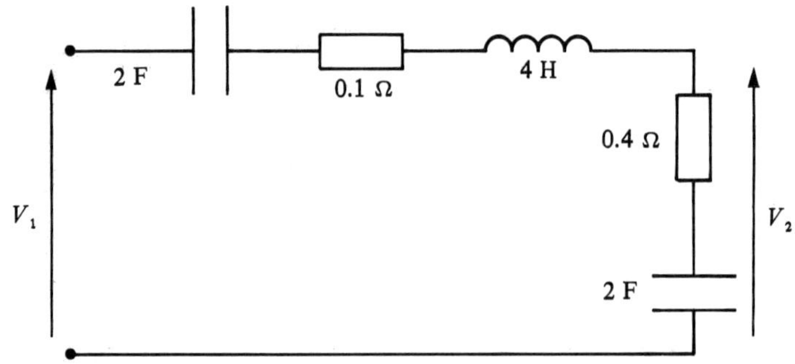

Fig. 4.3 The *LCR* circuit for Example 4.3.

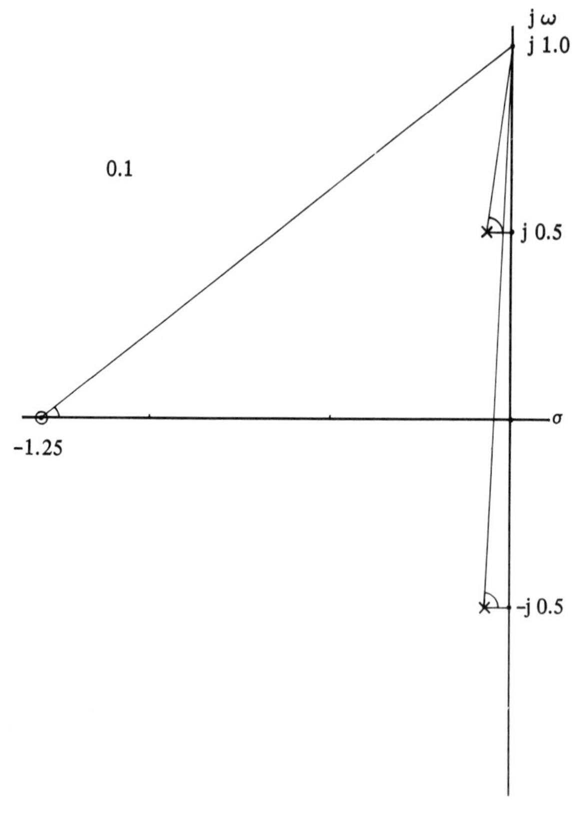

Fig. 4.4 The poles and zeros for the *LCR* circuit in Example 4.3; directed line segments for $T(\text{j}1)$.

Solution. The capacitors and inductors transform into $\frac{0.5}{s}$ and $4s$ respectively. The required transfer function is a simple potential divider relationship.

$$T(s) = \frac{V_2(s)}{V_1} = \frac{0.4 + 0.5/s}{4s + 0.5 + 1/s} . \qquad \ldots(4.30)$$

This result is rearranged and the quadratic is factorized to give:

$$T(s) = \frac{0.1\,(s + 1.25)}{(s + 0.0625 - \mathrm{j}0.5)\,(s + 0.0625 + \mathrm{j}0.5)} . \qquad \ldots(4.31)$$

The pole zero diagram for this transfer functiŏn is shown in Fig. 4.4. An investigation of the frequency response requires expression (4.28) to be applied for a range of values of ω. The directed line segments are shown in Fig. 4.4 for $\omega = 1$. Applying (4.28), the resulting $|T(\mathrm{j}\omega)|$ is given by:

$$|T(\mathrm{j}1)| = \frac{0.1 \times 1.6}{0.504 \times 1.501} = 0.21 . \qquad \ldots(4.32)$$

Other values of $|T(\mathrm{j}\omega)|$ can be found by using the same method. Note that the segments from the zero and from the pole at $-0.0625 - \mathrm{j}0.5$ will reduce in length at about the same rate as ω is reduced. The segment from the remaining pole at $-0.0625 + \mathrm{j}0.5$, however, will reduce rapidly as ω approaches 0.5 and then increase again for smaller values of ω. Thus we can expect to observe the resonance effect that can be seen in Fig. 4.5. This graph has been drawn by taking a range of values of ω between 0 and 1 and then applying (4.28) in each case. In order that the technique can be practised, the reader is advised to check the results at $\omega = 0.5$ and 0 respectively and find that $|T(\mathrm{j}\omega)|$ is 1.93 and 0.5.

The phase shift at a specified frequency is obtained by application of expression (4.29) to the pole zero diagram in Fig. 4.4. The result for $\omega = 1\,\mathrm{rad\,s^{-1}}$ is given by:

$$\underline{/T(\mathrm{j}1)} = 38.7° - 82.9° - 87.6° = -131.8° . \qquad \ldots(4.33)$$

Example 4.4. The voltage gain for a two stage *RC* coupled amplifier is given by the following expression:

$$A(\mathrm{j}\omega) = \frac{500}{\left(1 - \frac{\mathrm{j}10}{\omega}\right)\left(1 - \frac{\mathrm{j}40}{\omega}\right)\left(1 + \frac{\mathrm{j}\omega}{10^4}\right)\left(1 + \frac{\mathrm{j}\omega}{2 \times 10^4}\right)} . \qquad \ldots(4.34)$$

Determine the transfer function $A(s)$ and with the aid of a pole zero diagram, estimate the gain $A(\mathrm{j}\omega)$ at angular frequencies of 30, 500 and $10^4\,\mathrm{rad\,s^{-1}}$.

Solution. The required results could, of course be obtained by the substitution of values into expression (4.34). The object of this example is to gain further experience with the relationships between system transfer functions, the pole zero diagram and the steady-state response.

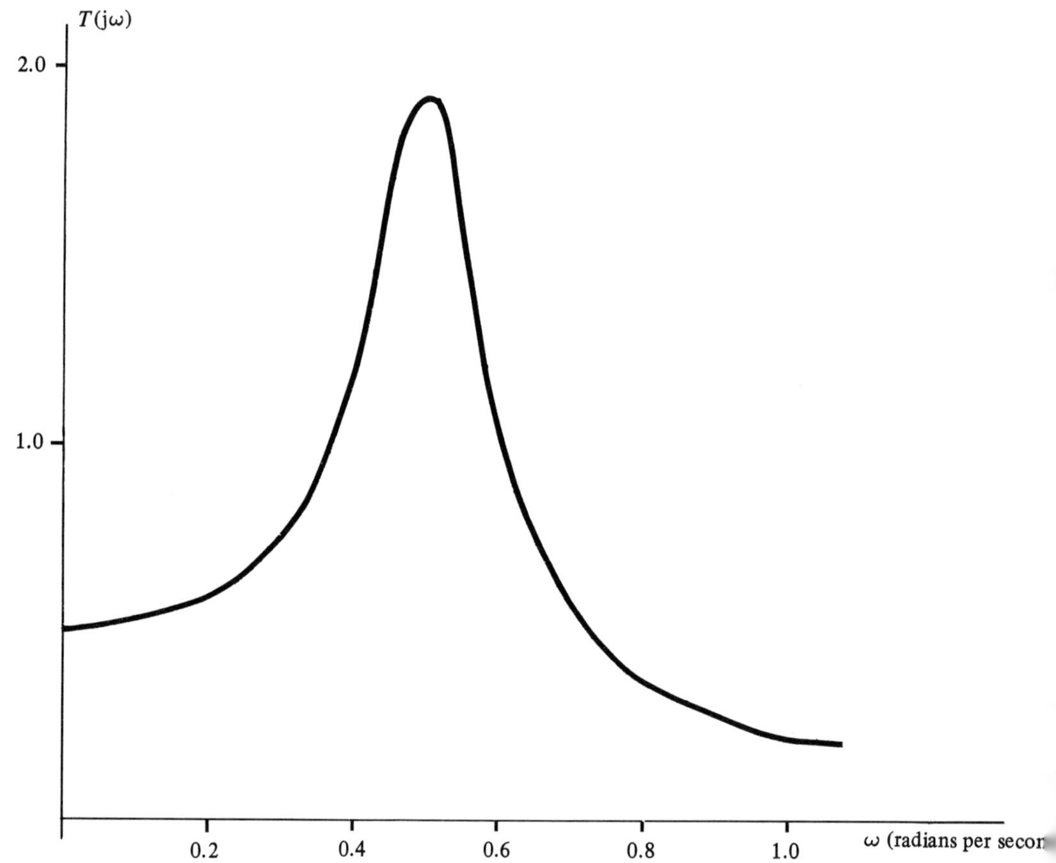

Fig. 4.5 The frequency response for $|T(j\omega)|$ for Example 4.3.

First, the terms $(1 - j\frac{10}{\omega})$ and $(1 - j\frac{40}{\omega})$ are changed to $(1 + \frac{10}{j\omega})$ and $(1 + \frac{40}{j\omega})$ respectively. $A(s)$ can then be written by substituting s for $j\omega$.

$$A(s) = \frac{500}{\left(1 + \frac{10}{s}\right)\left(1 + \frac{40}{s}\right)\left(1 + \frac{s}{10^4}\right)\left(1 + \frac{s}{2 \times 10^4}\right)} \quad \ldots(4.35)$$

Now multiply numerator and denominator by $2 \times 10^8 s^2$,

$$A(s) = \frac{10^{11} s^2}{(s + 10)(s + 40)(s + 10^4)(s + 2 \times 10^4)} \quad \ldots(4.36)$$

Drawing the pole zero diagram now presents some problems as linear scales are required, yet a wide range of values must be considered. The solution is to either draw two diagrams for different frequency ranges or to use broken scales as shown in Fig. 4.6.

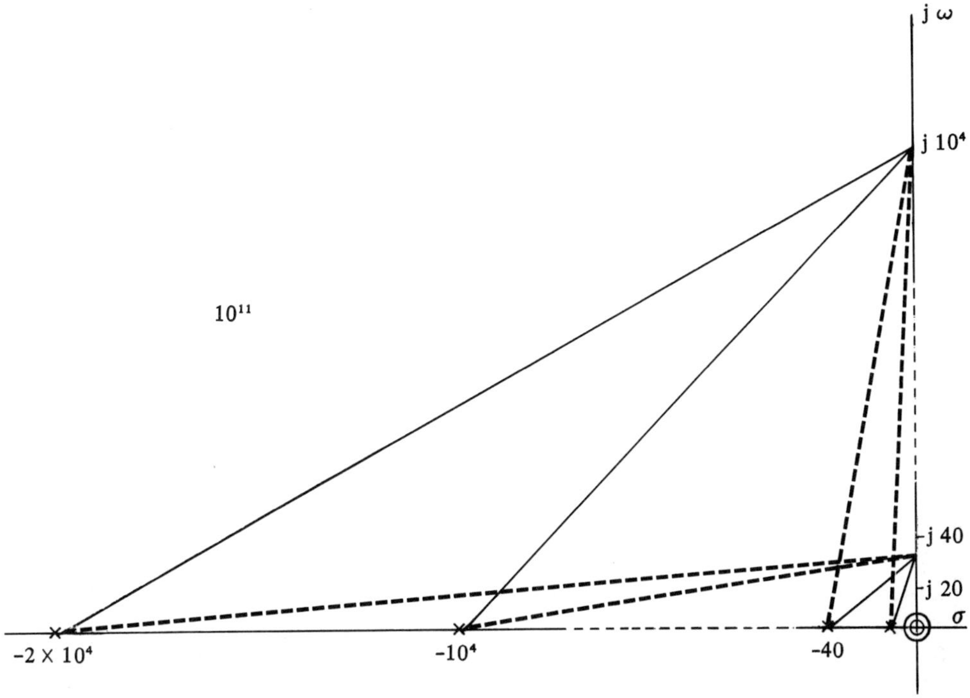

Fig. 4.6 The pole zero diagram for Example 4.4. (Note 'broken' scales.)

Consider the first frequency, $\omega = 30$ rads^{-1}; the directed line segments are shown from the two zeros and from the two low frequency poles. The lengths of these segments are respectively 30, 30, 31.6 and 50. The remaining two line segments from the high frequency poles are shown broken, as these are not to scale in either length or angle. If they were drawn to scale (to $j\omega = j30$), the lengths and angles would be approximately the same as the lengths and angles to the origin; i.e. $10^4\angle 0°$ and $2 \times 10^4\angle 0°$. Using these results;

$$|A(j30)| = \frac{10^{11} \times 30 \times 30}{31.6 \times 50 \times 10^4 \times 2 \times 10^4} = 285 \ . \qquad \ldots(4.37)$$

In a similar way, when $A(j10^4)$ is considered, the segments from the zeros and the two low frequency poles approximately cancel. The lengths from the two high frequency poles are $10^4\sqrt{2}$ and $10^4\sqrt{5}$ respectively. Thus, the required result is:

$$|A(j10^4)| = \frac{10^{11}}{10^4\sqrt{2} \times 10^4\sqrt{5}} = 316 \ . \qquad \ldots(4.38)$$

Finally, for the medium frequency, the low frequency poles again cancel with the zeros, but in this case, the segments from the high frequency poles are approximately the pole values of 10^4 and 2×10^4. Hence,

$$|A(j5000)| = \frac{10^{11}}{2 \times 10^4 \times 10^4} = 500 \ . \qquad \ldots(4.39)$$

This result is, of course, the original medium frequency gain.

4.5 THE POLE ZERO DIAGRAM AND THE BODE PLOT

Most readers will probably be familiar with the Bode plot representation of frequency response information. It will be worth recalling some of the important features of Bode plots before relating them to the pole zero diagram. Bode plots show the variation of |gain| and $\underline{/\text{gain}}$ with signal frequency. In the context of this book, these are $|T(\mathrm{j}\omega)|$ and $\underline{/T(\mathrm{j}\omega)}$ respectively.

The frequency scale is logarithmic which allows a wider range of frequency response information to be displayed. The gain scale is also logarithmic permitting a wide range and additionally allowing the responses to be represented by straight-line approximations. The logarithmic scale for the gain is the decibel scale where the number of decibels is given by $20 \log_{10}$ (gain) or in this case, $20 \log_{10} |T(\mathrm{j}\omega)|$. . . . (4.40)

Referring now to expression (4.20), which for convenience is repeated here:

$$|T(\mathrm{j}\omega)| = K \frac{|\mathrm{j}\omega + z_1| \cdot |\mathrm{j}\omega + z_2| \cdots |\mathrm{j}\omega + z_n|}{|\mathrm{j}\omega + p_1| \cdot |\mathrm{j}\omega + p_2| \cdots |\mathrm{j}\omega + p_m|} \qquad \ldots (4.20)$$

$|T(\mathrm{j}\omega)|$ is obtained from the product of a number of terms, so if logs are used for the individual terms: number of decibels = $20 \log_{10} |T(\mathrm{j}\omega)| = 20 \log_{10} K + 20 \log_{10} |\mathrm{j}\omega + z_1| + 20 \log_{10} |\mathrm{j}\omega + z_2| \ldots + 20 \log_{10} |\mathrm{j}\omega + z_n| - 20 \log_{10} |\mathrm{j}\omega + p_1| - 20 \log_{10} |\mathrm{j}\omega + p_2| - \ldots 20 \log_{10} |\mathrm{j}\omega + p_m|$ . . . (4.41)

Thus, if we can draw the Bode plot for each term separately, the results may be added to obtain the Bode gain plot for $|T(\mathrm{j}\omega)|$. Reference to expression (4.21) shows that the angle plot can be obtained in the same way, by addition, without using logs.

Each term to be added (for either plot), contains a single pole or a single zero. The first step then, is to consider the Bode plot resulting from a single pole or a single zero.

4.6 FREQUENCY RESPONSE RESULTING FROM SINGLE REAL POLES OR ZEROS

i. **A single real zero.** Suppose

$$T(s) = K(s + a) \qquad \ldots (4.42)$$

The pole zero diagram is shown in Fig. 4.7a. Now,

$$T(\mathrm{j}\omega) = K(\mathrm{j}\omega + a) = Ka\left(1 + \mathrm{j}\frac{\omega}{a}\right) \qquad \ldots (4.43)$$

which is the Bode form of $T(\mathrm{j}\omega)$.

Consider next the form of the frequency response resulting from (4.43).

If $\omega \ll a$, $\quad |T(\mathrm{j}\omega)| \triangleq Ka$, a constant. . . . (4.44)

If $\omega \gg a$, $\quad |T(\mathrm{j}\omega)| \triangleq K\omega$ and is directly proportional to ω. . . . (4.45)

These results can also be observed from the pole zero diagram; for very small values of ω, the length of the line segment is a, making $|T(\mathrm{j}\omega)| = Ka$; for very large ω, the length of the segment is ω, making $|T(\mathrm{j}\omega)| = K\omega$. These results can be seen on the log scale of the Bode gain plot in Fig. 4.7b.

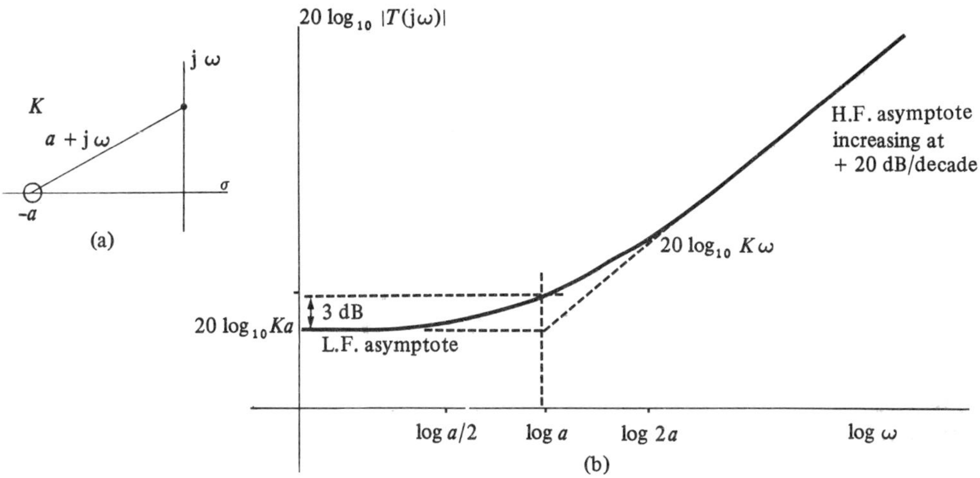

Fig. 4.7 The frequency response due to a single real zero. a) The pole zero diagram. b) The Bode gain plot.

For frequencies much less than a, the constant gain Ka appears as a horizontal line at $20\log_{10}Ka$ decibels. For frequencies much greater than a, an increase in frequency of a decade, results in an increase in $|T(j\omega)| = K\omega$ by a factor of 10, or for these scales, $20\log_{10}10$ or 20 dB. Thus the line will have a slope of 20 dB/decade (or 6 dB/octave). We can find the position of this line by noting that the *line* $|T(j\omega)| = K\omega$ coincides with the *line* $|T(j\omega)| = Ka$ at $\omega = a$.

The straight line approximation mentioned above includes only these two lines. For many purposes this will be sufficient, but at this stage we should examine the plot in more detail.

From (4.43), when $\omega = a$,

$$|T(j\omega)| = Ka(1 + j) \;, \qquad \ldots(4.46)$$

and

$$|T(j\omega)| = Ka\sqrt{2}\;.$$

In the logarithmic units,

$$20\log_{10}|T(j\omega)| + 20\log_{10}Ka + 20\log_{10}\sqrt{2} = 20\log_{10}Ka + 3 \text{ decibels}.$$

Similarly, at $\omega = a/2$ and $2a$ respectively, the points lie at 1 dB above the straight line approximation. Plotting these and other points results in the Bode gain plot shown. Note that if the straight line approximation is used, the greatest error is 3 dB and that this occurs at the zero frequency (or the pole frequency if a pole is being considered).

Now turning to the variation of $\underline{/T(j\omega)}$ with frequency. This can be obtained from expression (4.43) or by observation of the pole zero diagram in Fig. 4.7.

$$\underline{/T(j\omega)} = \tan^{-1}\omega/a \qquad \ldots(4.47)$$

This is plotted against $\log\omega$ in Fig. 4.8. A straight line approximation can be made by drawing a straight line from 0° at $\log a/10$ through 45° at $\log a$ to 90° at $\log 10a$. This approximation has a maximum error of less than 6°.

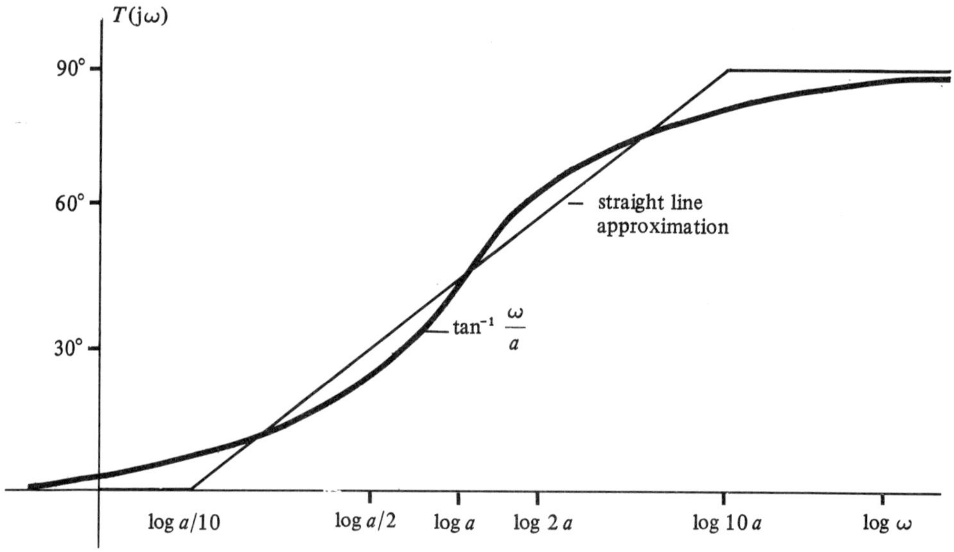

Fig. 4.8 The phase response due to a single real zero.

ii. A single real pole. Suppose

$$T(s) = \frac{K}{s+a}\,, \qquad \ldots(4.48)$$

then

$$T(j\omega) = \frac{K}{j\omega + a} = \frac{K/a}{1 + j\omega/a}\,. \qquad \ldots(4.49)$$

In this case, for $\omega \ll a$, $\quad T(j\omega) \simeq K/a\,, \qquad \ldots(4.50)$

and for $\omega \gg a$, $\quad T(j\omega) \simeq K/\omega\,. \qquad \ldots(4.51)$

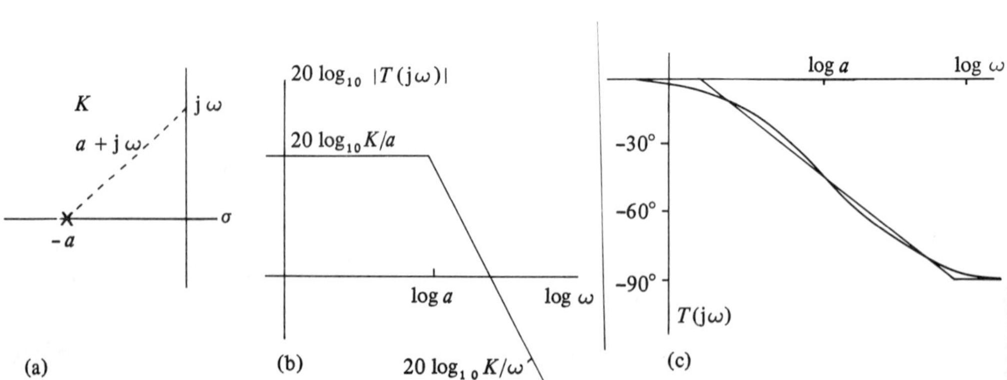

Fig. 4.9 The frequency response due to a single real pole. a) The pole zero diagram, b) The Bode gain plot. c) The Bode phase plot.

The pole zero diagram and the corresponding gain and angle Bode plots are shown in Fig. 4.9. It may be seen that the effect of the pole is exactly opposite to that of the zero. The results may be compared in the following table:

	$\omega < a$	$\omega = a$	$\omega > a$
Zero at $-a$	Gain constant. ϕ tends to 0°	Gain + 3 dB. $\phi = +45°$	Gain increases by 20 dB/decade or 6 dB/octave. ϕ tends to +90°
Pole at $-a$	Gain constant. ϕ tends to 0°	Gain − 3 dB. $\phi = -45°$	Gain falls by 20 dB/decade or 6 dB/octave. ϕ tends to −90°

4.7 FREQUENCY RESPONSE RESULTING FROM SEVERAL REAL POLES AND ZEROS

When a transfer function has several poles and zeros, the Bode plots for each can be drawn separately, and then added, to obtain the overall response. This can be best illustrated by an example.

Example 4.5. Construct the Bode gain and phase plots for the two systems having the following transfer functions.

a)
$$T_1(s) = \frac{2000\,(s+0.4)}{s\,(s+2)\,(s+10)\,(s+35)}\,. \qquad \ldots (4.52)$$

b)
$$T_2(s) = \frac{48 \times 10^{12} s^2}{(s+100)^2\,(s+10^5)\,(s+6\times 10^5)}\,. \qquad \ldots (4.53)$$

Consider alternative methods for the construction of the gain and phase plots and discuss the results with reference to the pole zero diagrams for the transfer functions.

Solution. a) The first step is to substitute for $s = j\omega$ and then to rearrange the result into the most convenient Bode form.

$$T_1(j\omega) = \frac{2000\,(j\omega+0.4)}{j\omega\,(j\omega+2)\,(j\omega+10)\,(j\omega+35)}\,. \qquad \ldots (4.54)$$

$$T_1(j\omega) = \frac{2000 \times 0.4\,(1+j\omega/0.4)}{2\times 10\times 35 j\omega\,(1+j\omega/2)\,(1+j\omega/10)\,(1+j\omega/35)}\,. \qquad \ldots (4.55)$$

$$T_1(j\omega) = \frac{1.14\,(1+j\omega/0.4)}{j\omega\,(1+j\omega/2)\,(1+j\omega/10)\,(1+j\omega/35)}\,. \qquad \ldots (4.56)$$

Note that each finite pole and zero have produced a term $(1 + j\omega/\omega_1)$, where $-\omega_1$ is the pole or zero frequency. The pole at the origin simply provides the denominator term $j\omega$. Each of the six terms in (4.56) can be plotted separately on the dB/log frequency scale shown in Fig. 4.10a. The constant 1.14 and the single $j\omega$ are conveniently taken together since at $\omega = 1$, $|1.14/j\omega| = 1.14$ which converts to 1.16 dB. Thus on the log scales, $1.14/j\omega$ appears as a straight line with a constant slope of –20 dB/decade passing through the point 1.16 dB at $\omega = 1$. The term associated with the zero, $(1 + j\omega/0.4)$, appears as 0 dB for ω less than 0.4 and then increases at +20 dB/decade for ω greater than 0.4. The terms resulting from the three poles; $(1 + j\omega/2)$, $(1 + j\omega/10)$ and $(1 + j\omega/35)$ each introduce 0 dB at frequencies below their pole or break frequency and for higher frequencies a fall at 20 dB/decade.

To obtain the total response, the five, individual, straight-line graphs are added. The most convenient frequencies for the addition are the break frequencies, as there will then only be straight lines between these points.

The construction of the phase response is shown in Fig. 4.10b. In this case, the constant 1.14 has no effect and the $1/j\omega$ from the pole at the origin introduces –90° at all frequencies. The phase shift due to the zero produces the straight-line approximation of 0° below 0.04 rads^{-1} (one tenth of the break frequency), an increase of 45°/decade up to 4 rads^{-1} (ten times the break frequency) and a constant +90° for frequencies greater than this. The similar straight-line approximations for the three poles each contribute 0° below one tenth of the pole frequency, then –45°/decade up to ten times the pole frequency and finally, –90° for all higher frequencies. The total is obtained, once again, by addition at points where one or more of the individual response terms change.

At this stage, it is worth comparing the main features of these Bode plots with the pole zero diagram in Fig. 4.10c. The following points may be observed:

i. The pole at the origin results in a low frequency asymptote increasing at 20 dB/decade to infinity at $\omega = 0$; it also causes a low frequency phase change of –90°.
ii. The final high frequency fall in gain is given by:
(Number of poles – number of zeros) × 20 dB/decade; in this example, 60 dB.
iii. The final high frequency phase shift is given by:
(Number of poles – number of zeros) × –90°; in this case, –270°.

Solution. b) Now turning to $T_2(s)$ which must be rearranged in the same way to obtain $T(j\omega)$.

$$T_2(j\omega) = \frac{(j\omega)^2 \times 0.08}{\left(1 + \frac{j\omega}{100}\right)^2 \left(1 + \frac{j\omega}{10^5}\right)\left(1 + \frac{j\omega}{6 \times 10^5}\right)} \quad \ldots (4.57)$$

A reference level for medium frequencies can first be determined. Then, instead of drawing all the individual terms and adding, we just draw the changes introduced by each term in turn. This is illustrated in Fig. 4.11. This method can be a little more difficult as in some cases the starting point is not obvious. For this example, in the frequency range 10^3 to 10^4 rads^{-1}, expression (4.57) approximates to:

$$\frac{(j\omega)^2 \times 0.08}{\left(\frac{j\omega}{100}\right)^2} = 800 \ .$$

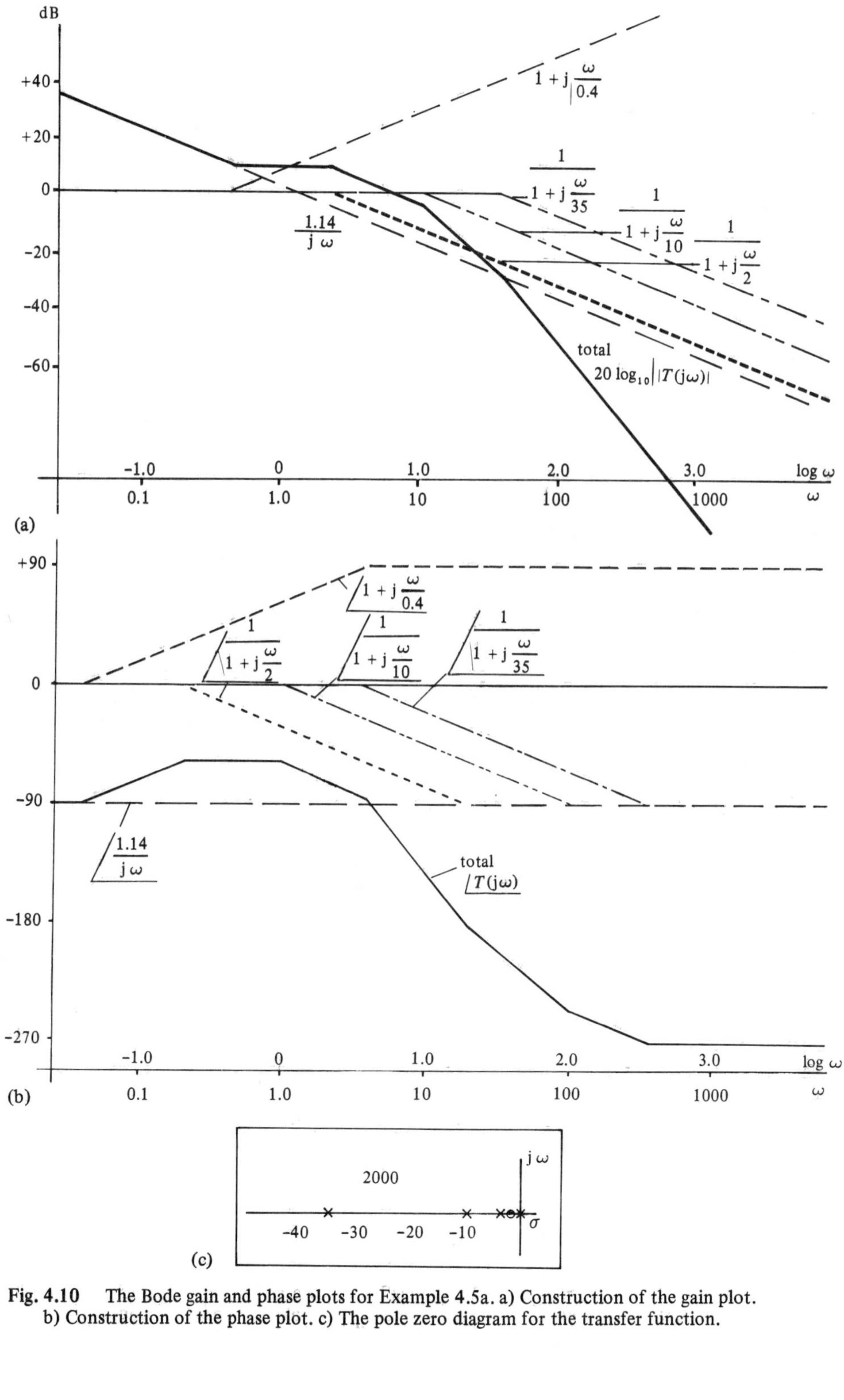

Fig. 4.10 The Bode gain and phase plots for Example 4.5a. a) Construction of the gain plot. b) Construction of the phase plot. c) The pole zero diagram for the transfer function.

Our reference level will therefore be 20 $\log_{10}$ 800, or 58 dB. This is drawn as a horizontal line over the frequency range mentioned.

At higher frequencies, this level continues up to $\omega = 10^5$ where the response starts to fall at 20 dB/decade due to the pole at that frequency. The fall continues up to $\omega = 6 \times 10^5$ where the second high frequency pole increases the slope to –40 dB/decade. Now turning to the low frequencies; a zero at the origin together with a low frequency pole produce a low frequency break causing the response to fall at 20 dB/decade as ω tends to zero. This is shown in Fig. 4.12.

In this problem, we have a double zero at the origin and a double pole at $\omega = -100$. This double break causes a fall of 40 dB/decade for ω below 100 rads^{-1}. No addition is required as the total response has been drawn in one procedure.

The phase response in Fig. 4.11b has been drawn in a similar way. In this case, starting at low frequencies, we find +180° from the double zero at the origin. As the frequency increases, the following changes apply:

i. From one tenth of each pole frequency, add –45°/decade to the slope.
ii. From ten times each pole frequency, add +45°/decade to the slope.
iii. From one tenth of each zero frequency, add +45°/decade to the slope.
iv. From ten times each zero frequency, add –45°/decade to the slope.

The reader is advised to apply the methods used for the construction of the responses of $T_2(j\omega)$ on $T_1(j\omega)$. (Start at a point well below the lowest finite pole and find a point on the $1.14/j\omega$ section.) Examination of the pole zero diagram in Fig. 4.11c confirms the points listed above (i to iv) and allows us to note some additional points.

v. A zero at the origin results in a low frequency asymptote decreasing by 20 dB/decade (6 dB/octave) to zero at $\omega = 0$. It also causes a low frequency phase change of +90°.
vi. The combination of a low frequency pole and a zero at the origin causes a break from +20 dB/decade (+6 dB/octave) (frequency increasing) to constant at the pole frequency.

4.8 FREQUENCY RESPONSE RESULTING FROM COMPLEX CONJUGATE POLES AND ZEROS

Pairs of complex conjugate poles are more common than zeros and these are therefore dealt with first.

Consider a transfer function;

$$T(s) = \frac{100s}{s^2 + 10s + 10^4} \,. \qquad \ldots (4.58)$$

This has a zero at the origin and a pair of complex conjugate poles at $-5 + j99.9$ and $-5 - j99.9$.

$$T(j\omega) = \frac{100\,j\omega}{(j\omega)^2 + 10\,j\omega + 10^4} \,. \qquad \ldots (4.59)$$

Extract $10j\omega$ as a factor from the denominator,

$$T(j\omega) = \frac{100j\omega}{10j\omega\left(\dfrac{j\omega}{10} + 1 + \dfrac{1000}{j\omega}\right)} = \frac{10}{\dfrac{j\omega}{10} + 1 + \dfrac{1000}{j\omega}} \,. \qquad \ldots (4.60)$$

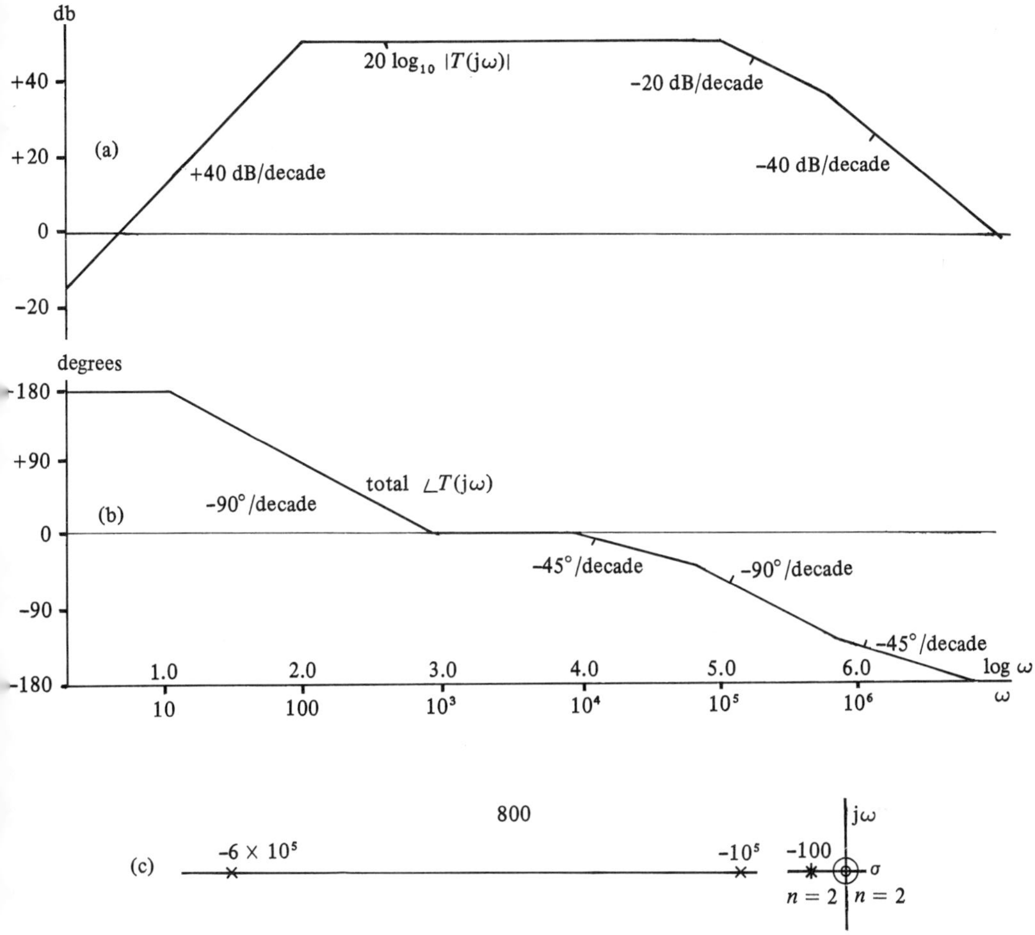

Fig. 4.11 The Bode gain and phase plots for Example 4.5b. a) Construction of the gain plot. b) Construction of the phase plot. c) The pole zero diagram for the transfer function.

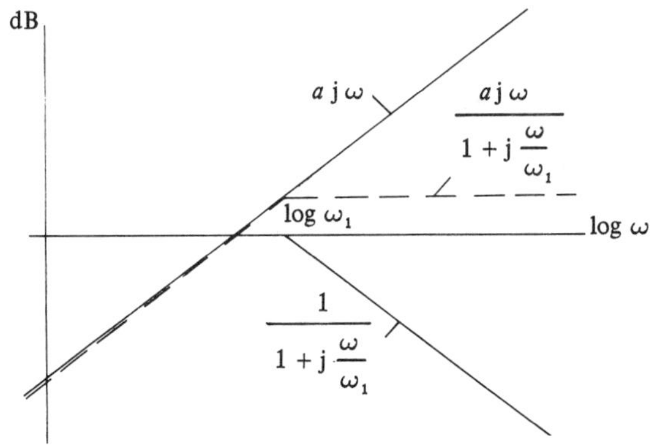

Fig. 4.12 The Bode gain plot resulting from a zero at the origin *and* a real pole.

The low and high frequency asymptotes for this expression can be found by taking the denominator in two parts, $\left(1+\frac{j\omega}{10}\right)$ giving the high frequency asymptote and $\left(1+\frac{1000}{j\omega}\right)$ for the low frequency asymptote. These are valid as, for high frequencies, $\frac{1000}{j\omega} \ll \frac{j\omega}{10}$, while at low frequencies the opposite is true. The term $\frac{1}{1+\frac{j\omega}{10}}$ produces 0 dB up to $\omega = 10$ and then –20 dB/decade. The term $\frac{1}{1+\frac{1000}{j\omega}}$ similarly introduces 0 dB *above* $\omega = 1000$ and then –20 dB/decade as ω falls below 1000 rads^{-1}. These two components are shown in Fig. 4.13. They are only valid for the whole expression (4.60), however, for the high and low frequency asymptotes.

Another point on the response due to the denominator can be obtained for the frequency at which $\frac{\omega}{10} = \frac{1000}{\omega}$. This frequency is given by $= \sqrt{10 \times 1000} = 100$. At $\omega = 100$, the denominator is $j10 + 1 - j10$ or 1. On the logarithmic scale this is 0 dB and the point is shown on the diagram. The move from the asymptotes to this point can be found by one or two spot calculations. For example, at $\omega = 50$ rads^{-1},

$$\frac{j\omega}{10} + 1 + \frac{1000}{j\omega} = j5 + 1 - j20 = 1 - j15,$$

$$\therefore \text{ denominator } = \sqrt{226} \text{ or } 23.5 \text{ decibels.}$$

Points can therefore be drawn on the graph at –23 dB for $\omega = 50$ and also at $\omega = 200$ as the curve is symmetrical.

Similarly, at $\omega = 80$, we have $j8 + 1 - j12.5$ which results in –13.2 dB and at $\omega = 90$ in –7.4 dB. The 'resonant' curve due to the denominator can now be drawn as shown. The overall response is now obtained by adding the 'resonant term' to any remaining terms, in this case the constant 20 dB from the numerator constant of 10.

The response in the above example is very selective as a result of the relatively small real part of the complex conjugate poles at $s = -5 + j99.9$ (see the pole zero diagram in Fig. 4.13b). For comparison purposes, Figs. 4.14 and 4.15 show the construction for less selective transfer functions having the same resonant frequency. In the first case, the two asymptotes break at the same point, $\omega = 100$, and in the second the breaks are such that the asymptotes no longer cross. Note that in the second case the true curve now lies below the asymptotes.

Another situation that is often encountered is a transfer function of the same form as those discussed above, but without the zero at the origin. An example can be used to illustrate this point.

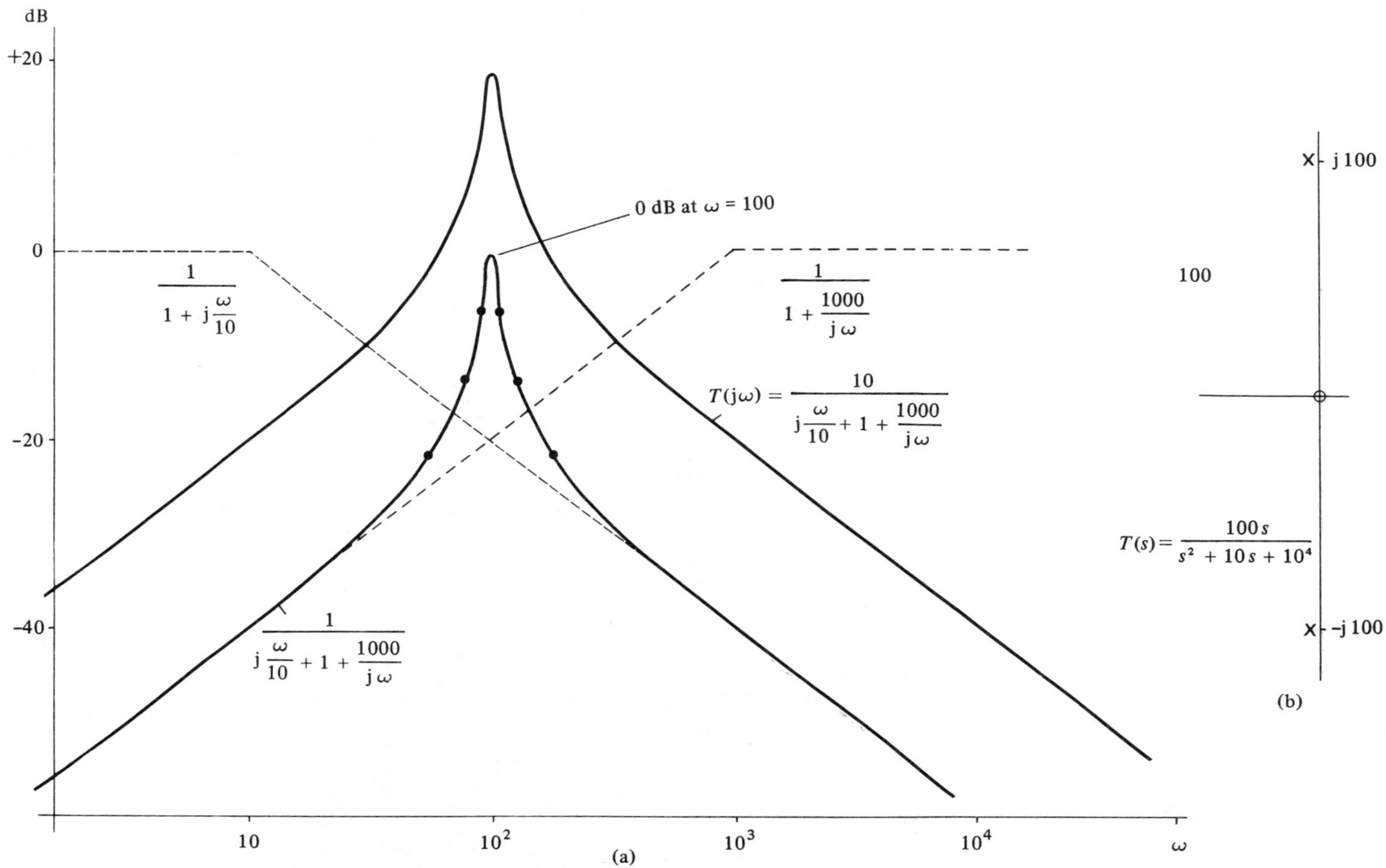

Fig. 4.13 The Bode gain plot for a transfer function having a pair of complex conjugate poles ($Q = 10$) and a zero at the origin. a) Construction. b) The pole zero diagram.

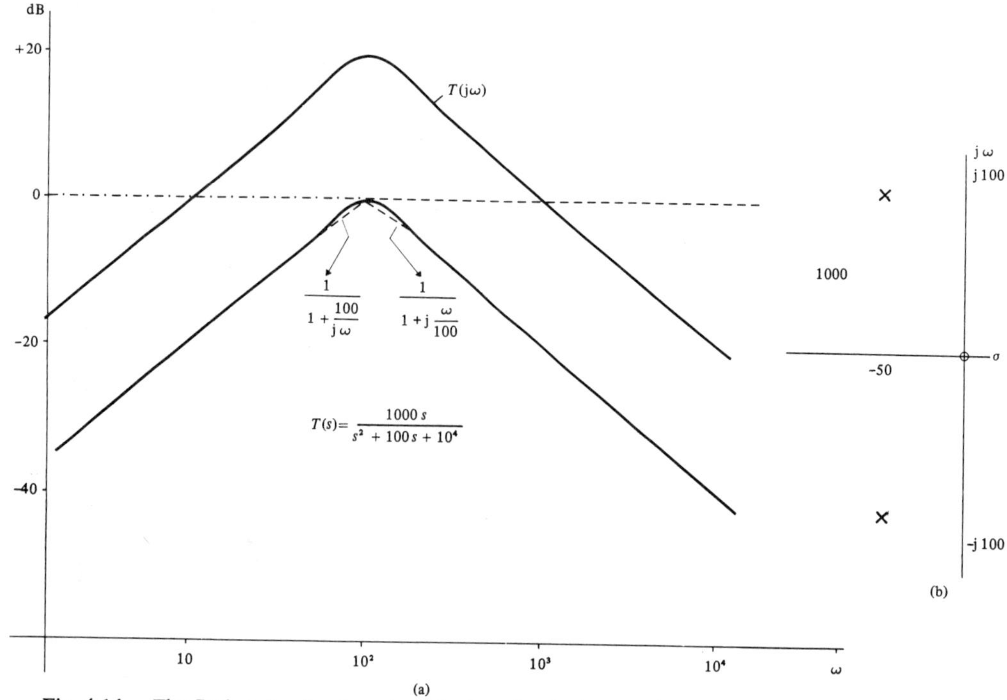

Fig. 4.14 The Bode gain plot for a transfer function having a pair of complex conjugate poles (Q = 1) and a zero at the origin. a) Construction. b) The pole zero diagram.

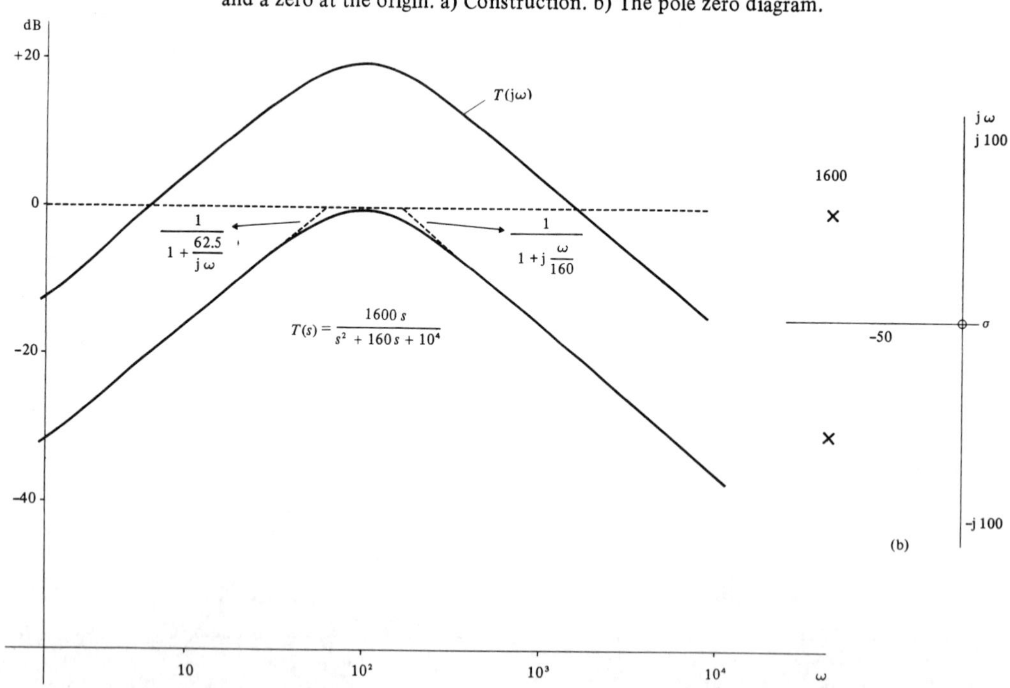

Fig. 4.15 The Bode gain plot for a transfer function having a pair of complex conjugate poles (Q = 0.625) and a zero at the origin. a) Construction. b) The pole zero diagram.

Example 4.6. The transfer function for a third-order Butterworth low pass filter is given by:

$$T(s) = \frac{125 \times 10^9}{(s + 5000)(s^2 + 5000s + 2.5 \times 10^7)} . \qquad \ldots (4.61)$$

Construct the Bode gain response for this filter.

Solution. The normal rearrangement to the Bode form for $T(j\omega)$ is left to the reader, but the following result should be obtained.

$$T(j\omega) = \frac{5000}{j\omega \left(1 + \frac{j\omega}{5000}\right)\left(\frac{j\omega}{5000} + 1 + \frac{5000}{j\omega}\right)} . \qquad \ldots (4.62)$$

The three components of this expression, $\frac{5000}{j\omega}$, $\frac{1}{\left(1 + \frac{j\omega}{5000}\right)}$ and $\frac{1}{\left(\frac{j\omega}{5000} + 1 + \frac{5000}{j\omega}\right)}$ are shown separately in Fig. 4.16, together with the over-all result. Note that at low frequencies, the $\frac{5000}{j\omega}$ term cancels the negative asymptote of the quadratic term to give a constant of 0 dB for this frequency range. Nearer the cut-off frequency of 5000 rads^{-1}, the rise above the asymptote combines with the simple $\frac{1}{1 + \frac{j\omega}{5000}}$ and the $\frac{5000}{j\omega}$ to maintain the zero decibels up to just below cut-off. At cut-off, the normal –3 dB applies. At higher frequencies, the three asymptotes combine to give a roll-off of –60 dB/decade.

4.9 *Q* FACTOR AND RESONANT RISE

In the last example, there was no resonant rise as the two asymptotes did not cross as they did in the example before that shown in Fig. 4.13. This condition relates with the angle that the complex poles make with the negative real axis. This is shown in Fig. 4.16b and is 60°. An angle greater than 60° results in a resonant rise or a *Q* factor of greater than unity.

For most readers, the *Q* factor will be most familiar for the simple series tuned circuit and it is therefore convenient to use this circuit to derive a general relationship between *Q* and the angle of the complex poles.

Consider a series *LCR* circuit; the current is given by:

$$I(s) = \frac{V(s)}{sL + R + 1/sC} , \qquad \ldots (4.63)$$

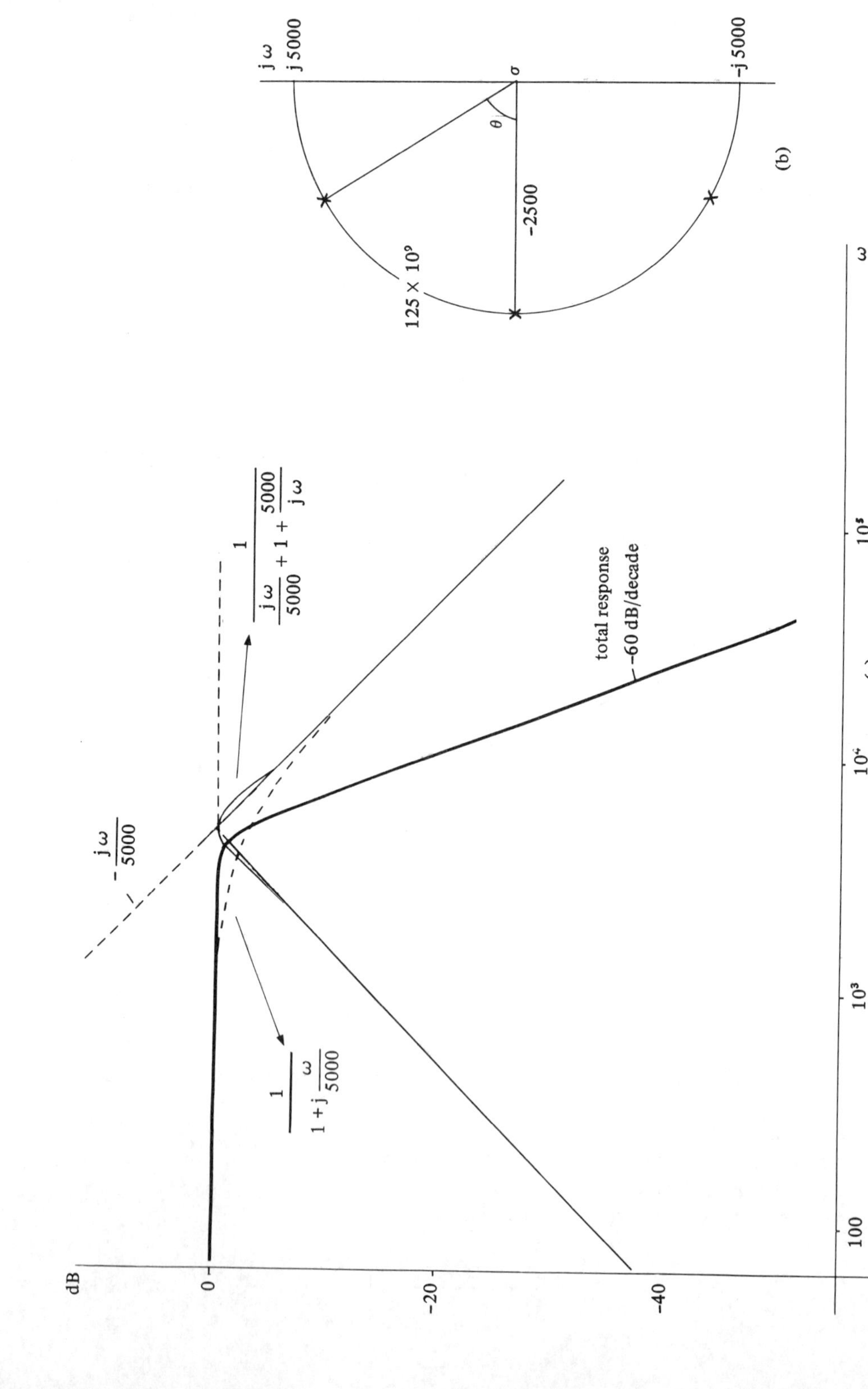

Fig. 4.16 The Bode gain response for a third-order Butterworth low pass filter. a) Construction. b) The pole zero diagram.

$$= \frac{sV(s)}{s^2 + \frac{R}{L^s} + \frac{1}{LC}} . \qquad \ldots (4.64)$$

The poles for this expression lie at:

$$s = -\frac{R}{2L} \pm j\sqrt{\frac{1}{LC} - \frac{R^2}{4L^2}} \qquad \ldots (4.65)$$

The pole zero diagram for expression (4.64) is shown in Fig. 4.17.

$$\cos\theta = \frac{R/2L}{\sqrt{\frac{1}{LC} - \frac{R^2}{4L^2} + \frac{R^2}{4L^2}}} = \frac{R}{2}\sqrt{\frac{C}{L}} . \qquad \ldots (4.66)$$

But,

$$Q = \frac{1}{R}\sqrt{\frac{L}{C}} . \quad \text{(See p. 88)} \qquad \ldots (4.67)$$

$$\therefore \quad \frac{1}{\cos\theta} = \frac{2}{R}\sqrt{\frac{L}{C}} = 2Q , \qquad \ldots (4.68)$$

$$\therefore \quad Q = \frac{1}{2\cos\theta} . \qquad \ldots (4.69)$$

(Compare this result with the section in Chapter 3 (page 88) on damping factor ζ.)

The Q factor can now be related to the sketch of a resonant response shown in Fig. 4.18. Now, for a high Q value, θ tends to 90° and the imaginary part of the pole frequency tends to ω_0. Also, the bandwidth $\frac{\omega_0}{Q}$ becomes smaller and the function is more frequency selective.

When $\theta = 60°$, $Q = 1$, the bandwidth is ω_0 and there is no resonant rise. Note that although there is no resonant rise if θ is less than 60°, the transient response still includes a decaying sinusoid as long as the poles are complex (i.e. if θ is greater than 0°).

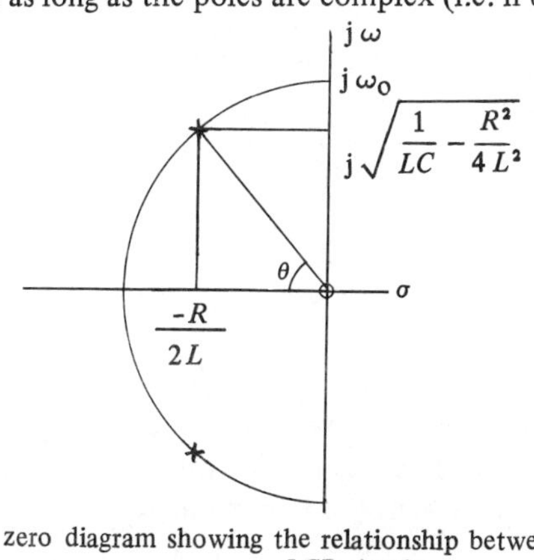

Fig. 4.17 The pole zero diagram showing the relationship between Q factor and pole position for an *LCR* circuit.

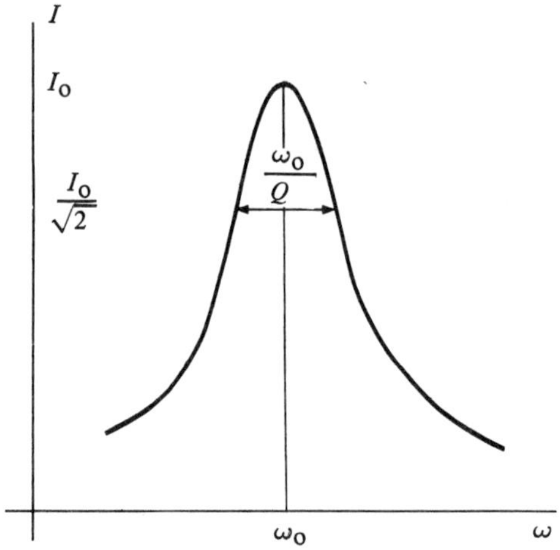

Fig. 4.18 The frequency response for an *LCR* circuit showing the relationship between Q factor, resonant frequency ω_0 and bandwidth.

4.10 COMPLEX CONJUGATE ZEROS

The effect of a pair of complex conjugate zeros is the exact inverse of that due to the complex poles. A final example in this section illustrates the effect of such zeros and that due to a high Q denominator with no zero at the origin.

Example 4.7. A system has a transfer function $T(s)$ given by:

$$T(s) = \frac{s^2 + 4s + 29}{s^2 + 2s + 82} . \qquad \ldots (4.70)$$

Draw the pole zero diagram representing this transfer function and compare this with a construction of the Bode gain plot.

Solution. Solving the two quadratics shows that the poles of $T(s)$ are at $-1 + j9$ and $-1 - j9$ while the zeros are at $-2 + j5$ and $-2 - j5$. These are shown in the pole zero diagram in Fig. 4.19b.

Substituting $j\omega$ for s in expression (4.70) and rearranging as with previous examples, results in:

$$T(j\omega) = \frac{2\left(\frac{j\omega}{4} + 1 + \frac{7.25}{j\omega}\right)}{\left(\frac{j\omega}{2} + 1 + \frac{41}{j\omega}\right)} . \qquad \ldots (4.71)$$

In Fig. 4.19a, the numerator and denominator terms are constructed separately using the methods described above. For the numerator expression, the two asymptotes are increasing at 20 dB/decade instead of the fall of 20 dB/decade of those due to the denominator term. The final response is obtained as before using point by point addition.

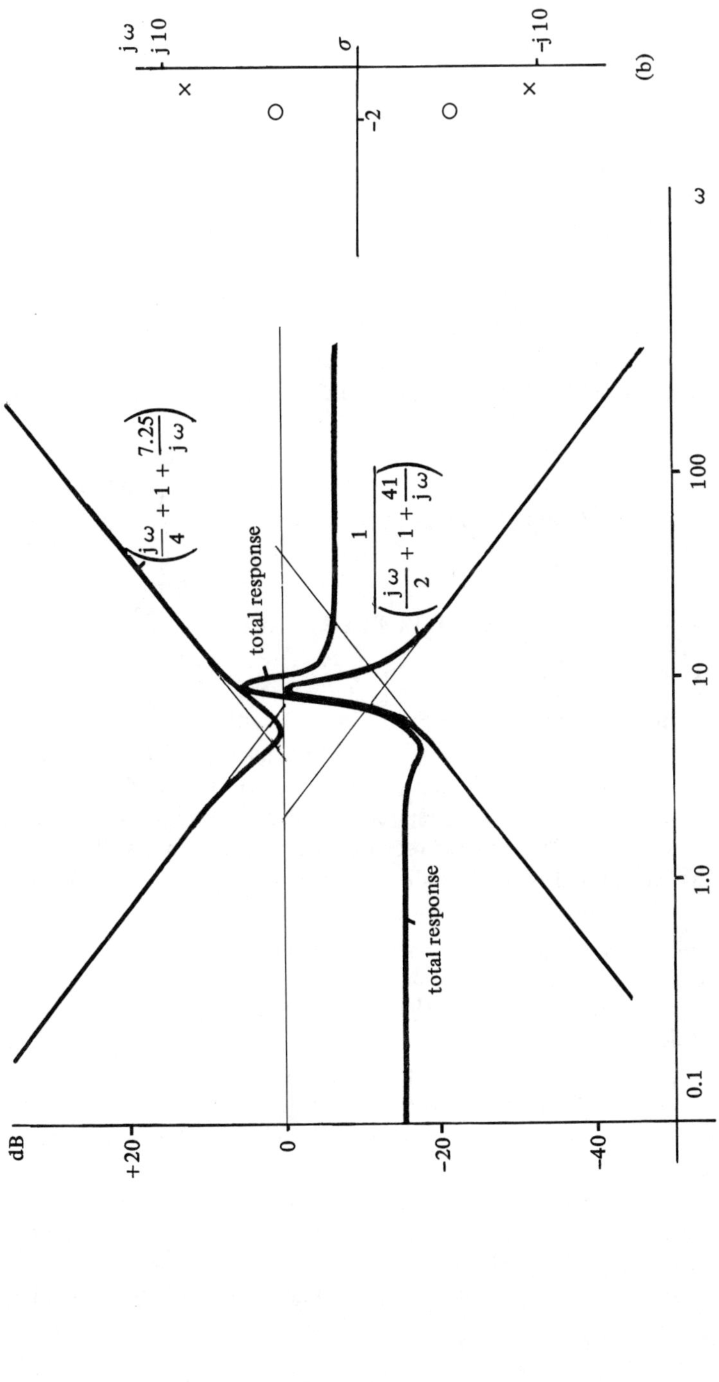

Fig. 4.19 The Bode gain plot for a system having complex poles and complex zeros. a) Construction. b) The pole zero diagram.

Comparison with the pole zero diagram shows that, while the poles correspond to a resonant rise at a frequency of 9 rads^{-1} (the imaginary part of the poles) the zeros correspond to a resonant fall or dip at 5.4 rads^{-1}, which is near to the value of the imaginary part of the zeros. The angle the zeros make with the real axis is smaller than that to the poles. The response is therefore less selective.

4.11 GENERAL DISCUSSION ON POLES AND ZEROS AND THE STEADY-STATE RESPONSE

In the preceding sections of this chapter, the construction techniques for the plotting of Bode gain and phase plots have been examined in detail. In an earlier section, the effects of poles and zeros upon the response at single frequencies were discussed. The remainder of this chapter considers a number of guidelines by which the pole zero diagram can be 'read' so that the steady-state frequency response can be predicted in general terms.

The pole zero diagrams used to illustrate this section are not strictly to scale as, in some cases, only the use of logarithmic scales could prevent the loss of detail at low frequencies. Such scales cannot be used as the line segment relationships would no longer be valid. Where a change in scale is required, this is indicated by a break in the axis. Wherever practical, examples of circuits or systems without values are included. Such examples are, however, not the only systems which could be used to obtain the particular pole zero diagram; i.e. a pole zero diagram does not imply a unique circuit or system.

4.12 SYSTEMS WITH REAL NEGATIVE POLES ONLY

Three pole zero diagrams are shown in Fig. 4.20, each with the associated Bode plots and a circuit which could produce the pattern shown. Diagrams (a) and (b) each have three poles but with (a) they are all spaced well apart, while with (b) the two high frequency poles are relatively close together. In both cases, the resulting gain response is constant at low frequencies; it then falls at 20 dB/decade for frequencies above that of the first pole, 40 dB/decade after the second and 60 dB/decade after the third. Thus the final high frequency fall in gain is N_p(20 dB/decade), where N_p is the number of poles.

The maximum error in the straight-line approximation occurs at the pole frequency where the true plot lies 3 dB below the one shown. The phase plot shows a smooth change for 0° allow frequencies to –270° at high frequencies. If the spacing between the poles is more than a decade, the change is not smooth as shown in case (b). The final phase lag is in any case, $N_p \times 90°$. With case (b), the gain plot shows the change of slope from 20–40 dB/decade and 40–60 dB/decade almost coincide (as they would if the two poles were coincident). The maximum error in the straight-line graph will now approach 6 dB in the region of the pole frequencies.

The circuit for (c) includes an integrator which provides the pole at the origin. The gain plot for this starts at infinity when ω is zero and falls continually at 20 dB/decade. The finite pole then increases the slope to 40 dB/decade. The phase angle here will be –90° for all low frequencies and it will then fall to –180° at high frequencies ($N_p \times 90°$).

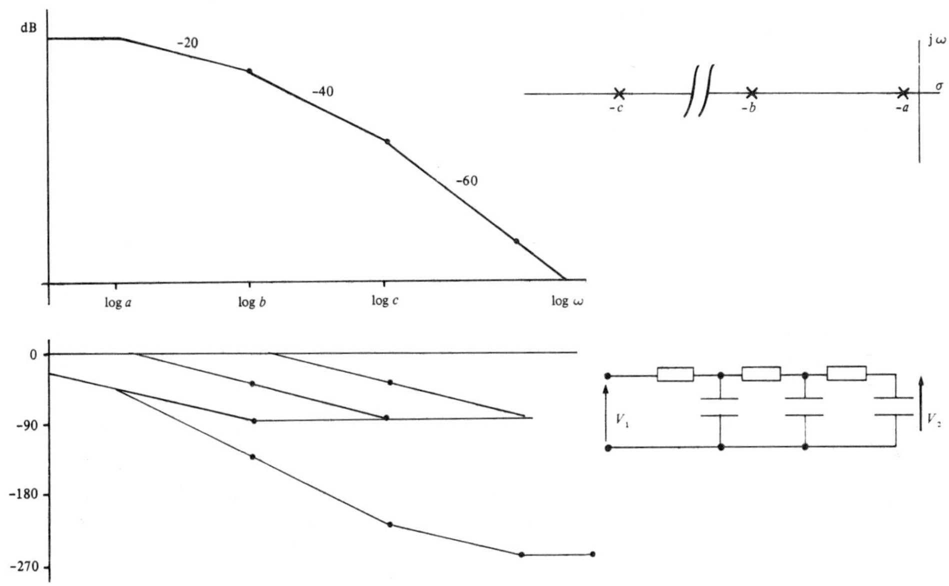
dB
-20
-40
-60
log a
log b
log c
log ω
jω
σ
-c
-b
-a
0
-90
-180
-270
V_1
V_2

(a)

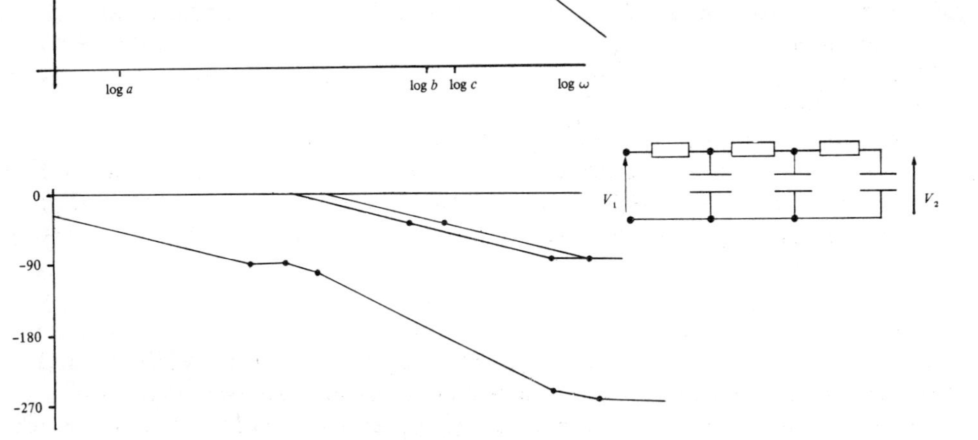
dB
-20
-40
-60
log a
log b
log c
log ω
jω
σ
0
-90
-180
-270
V_1
V_2

(b)

Fig. 4.20

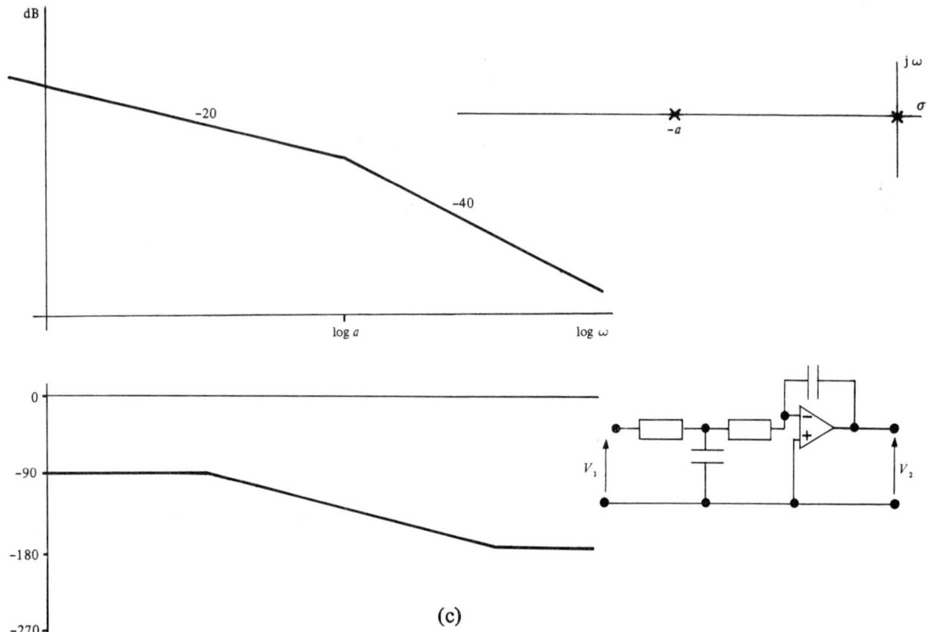

(c)

Fig. 4.20 Bode plots and pole zero diagrams for systems having only real negative poles.

4.13 SYSTEMS WITH REAL NEGATIVE POLES AND ZEROS

Three examples in this class have been shown in Fig. 4.21. Circuit (a) represents an *RC* coupled amplifier in which the two series capacitors produce the two zeros at the origin. These make the low frequency gain response rise at 40 dB/decade from zero at d.c. The first two poles reduce this to 20 dB/decade and constant respectively, the last pole then producing a 20 dB/decade fall. The general rule for the high frequency response is $(N_p - N_z)$ (20 dB/decade). For the low frequency response, any zeros at the origin result in a zero d.c. gain and the low frequency slope is (number of zeros at the origin) (+20 dB/decade).

The phase plot shows +180° at low frequencies followed by a fairly smooth change to −90° at high frequencies. If poles (b) and (c) were further apart, there would be a range of ω for which there would be no phase change.

Circuit (b) shows how the effects of poles and zeros can be taken in turn to give positive slopes, constants and negative slopes as frequency increases. The general rules regarding high and low frequency responses outlined above should be checked against this result. If a pole and zero are close together (less than 2:1) their effects nearly cancel and they can be ignored in sketching a response.

Circuit (c) is similar, but in this case, it is the practical arrangement for a lag-lead compensator used in feedback and control systems. The phase plot shows the low frequency pole and zero result in a frequency range with a lagging phase angle, while at a higher frequency range, the remaining zero and pole cause a leading angle. With typical spacing of the poles and zeros, the maximum lag and lead are respectively about 55°. The gain response is constant at both low and high frequencies but a fall in gain in the lag-lead region is demonstrated.

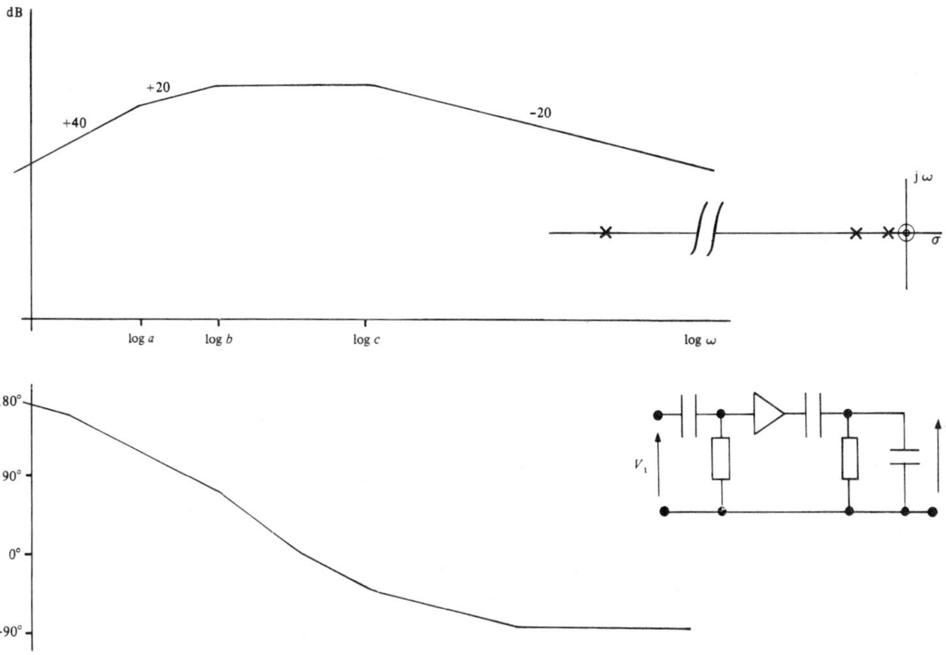
dB
+40
+20
-20
log a
log b
log c
log ω
j ω
σ
+180°
+90°
0°
-90°
V₁
V₂

(a)

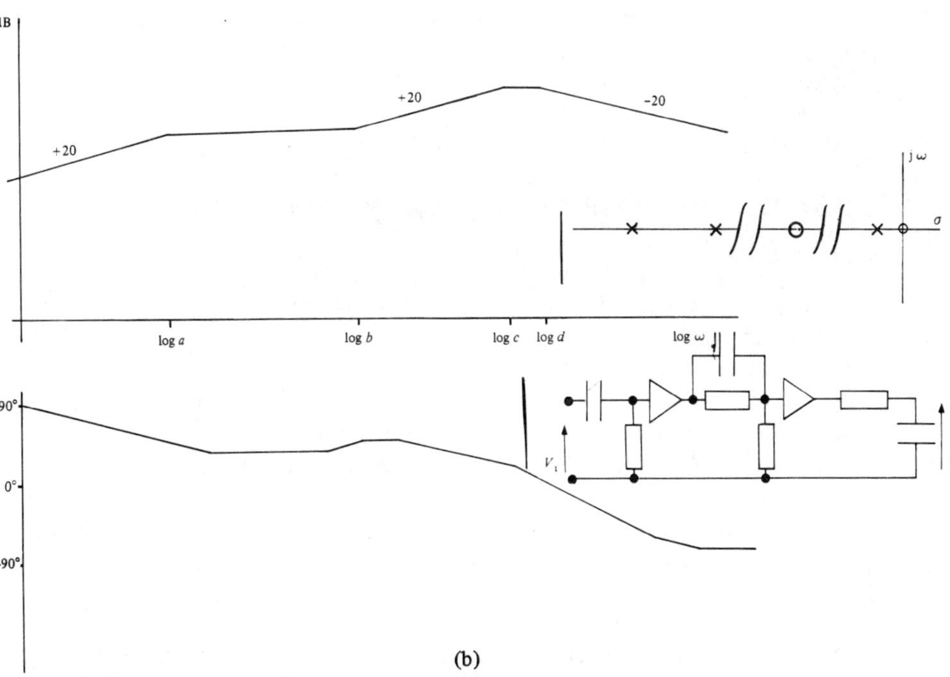
dB
+20
+20
-20
log a
log b
log c
log d
log ω
j ω
σ
+90°
0°
-90°
V₁

(b)

Fig. 4.21

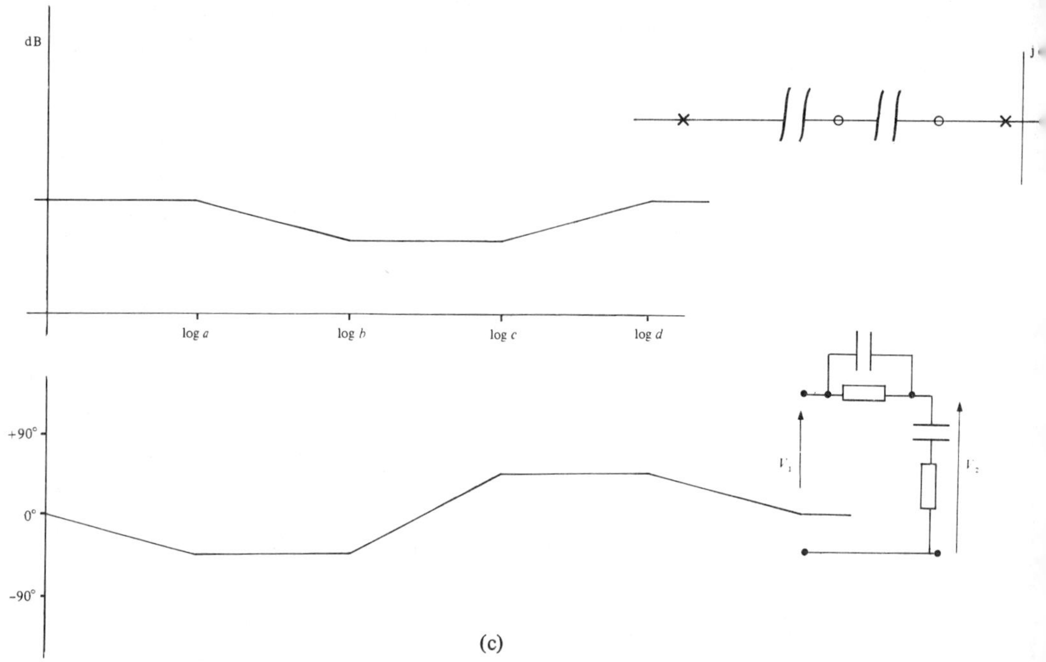

(c)

Fig. 4.21 Bode plots and pole zero diagrams for systems having real negative poles and zeros.

4.14 SYSTEMS WITH A COMPLEX CONJUGATE PAIR OF POLES

Systems in this class will either involve two types of energy storage, such as in an *LCR* circuit or with a pendulum, or amplification and feedback such as occurs in feedback amplifiers and control systems.

The first example in Fig. 4.22a has a complex conjugate pair of poles at an angle θ with the real axis. The general rules for poles and zeros still apply, i.e. the gain and phase at very low frequencies will be zero and +90° respectively, as there is a zero at the origin. The high frequency response falls at 20 dB/decade and the phase tends to –90° as the system has two poles and one zero. The poles are shown on a semicircle of radius *a*, which means that regardless of θ, the maximum value of the transfer function $T(j\omega)$ will occur at $\omega = a$. The Q factor is given by $\dfrac{1}{2\cos\theta}$ and determines the bandwidth of the response at 3 dB below the maximum value. The various frequency responses show the effect of varying the angle θ and hence, the Q factor.

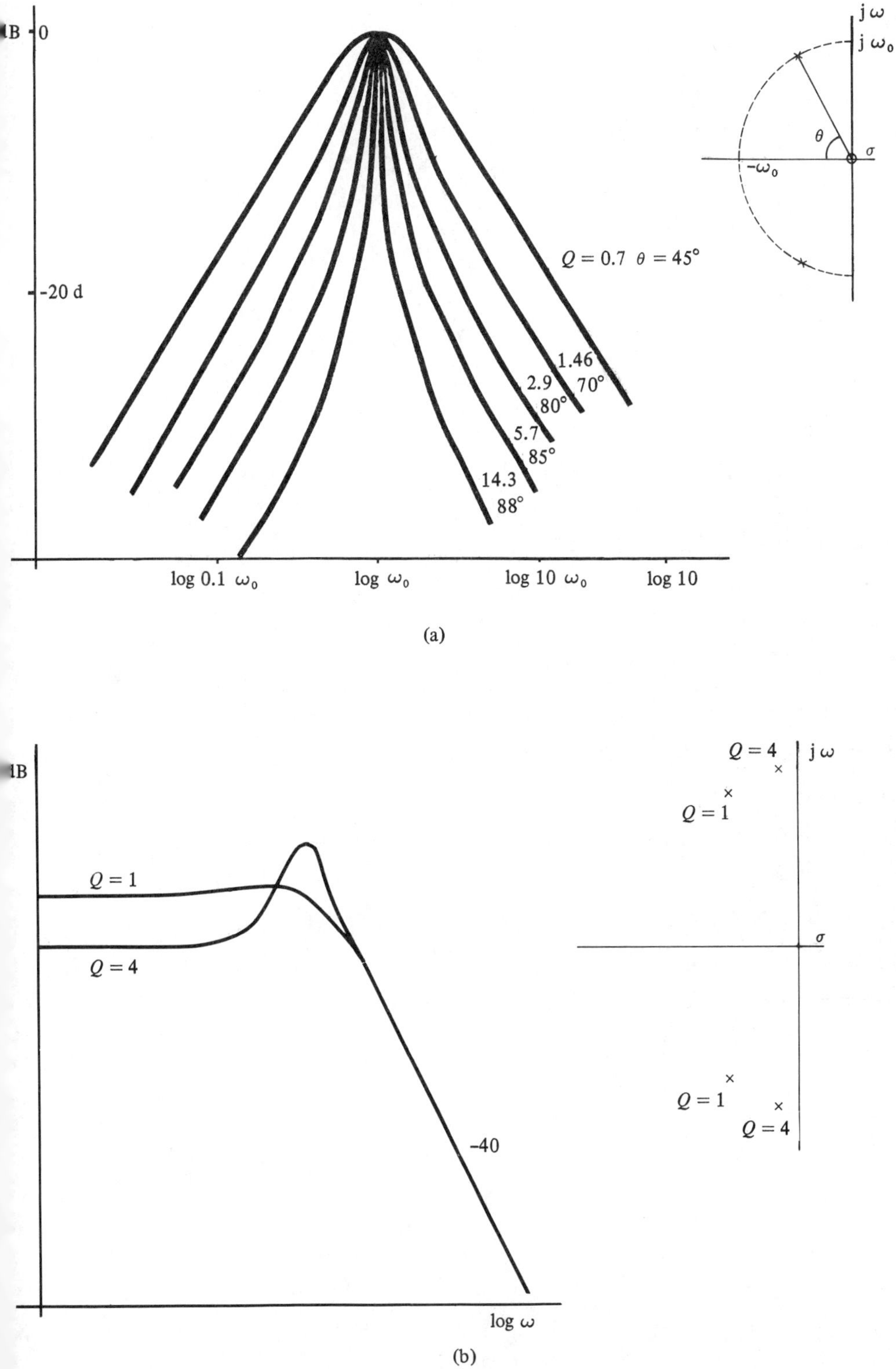
dB
0
-20 d
$Q = 0.7$ $\theta = 45°$
1.46
70°
2.9
80°
5.7
85°
14.3
88°
log 0.1 ω_0
log ω_0
log 10 ω_0
log 10
jω
jω_0
θ
σ
$-\omega_0$
(a)
dB
$Q = 1$
$Q = 4$
-40
log ω
$Q = 4$
jω
$Q = 1$
σ
$Q = 1$
$Q = 4$
(b)

Fig. 4.22

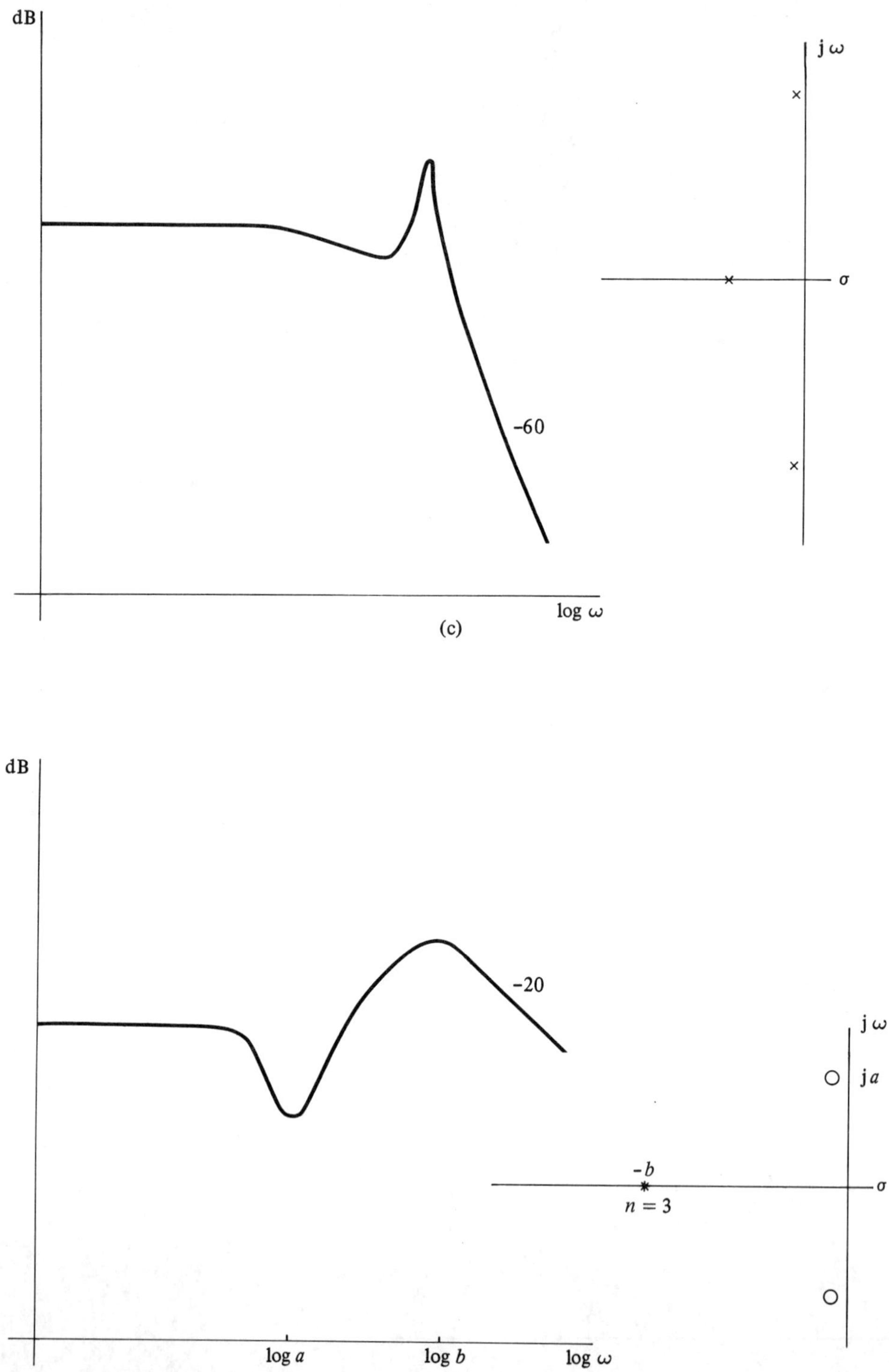

Fig. 4.22

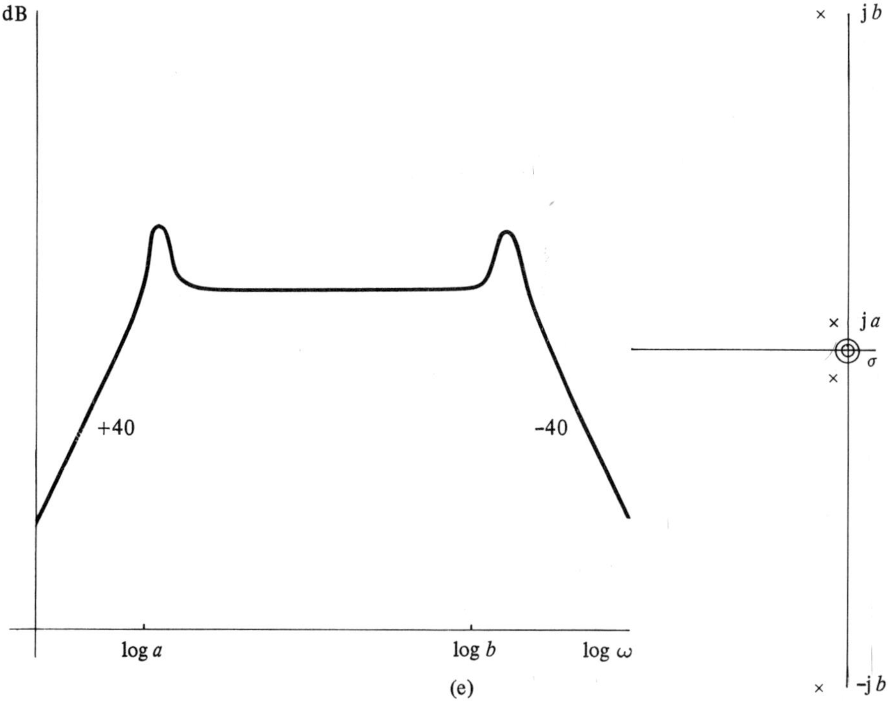
dB
+40
-40
log a
log b
log ω
j b
j a
σ
-j b

(e)

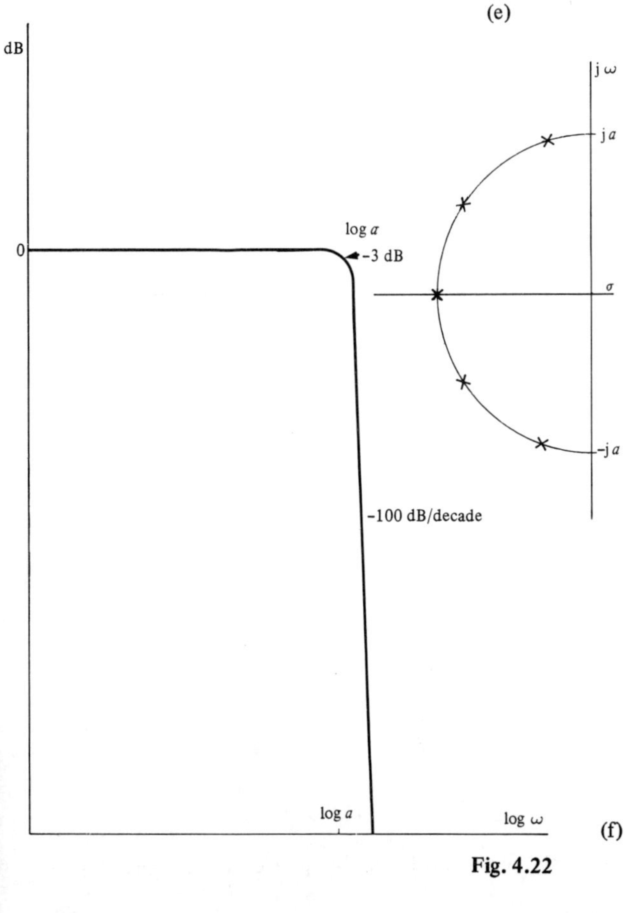
dB
0
log a
-3 dB
-100 dB/decade
log a
log ω
j ω
j a
σ
-j a

(f)

Fig. 4.22

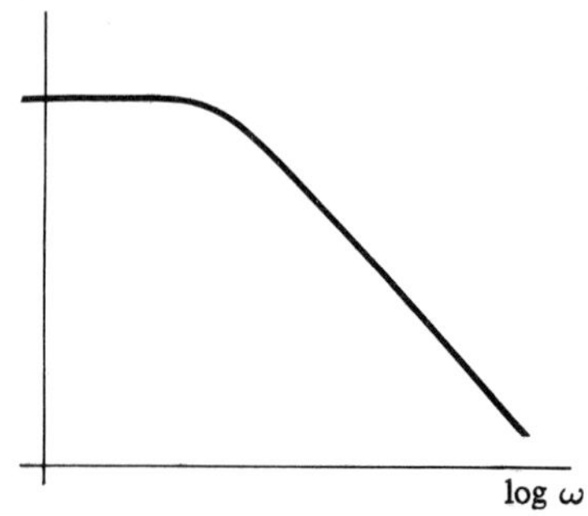

$$T(s) = \frac{1}{s^2 + \dfrac{\omega_p}{Q_p}s + \omega_p{}^2}$$

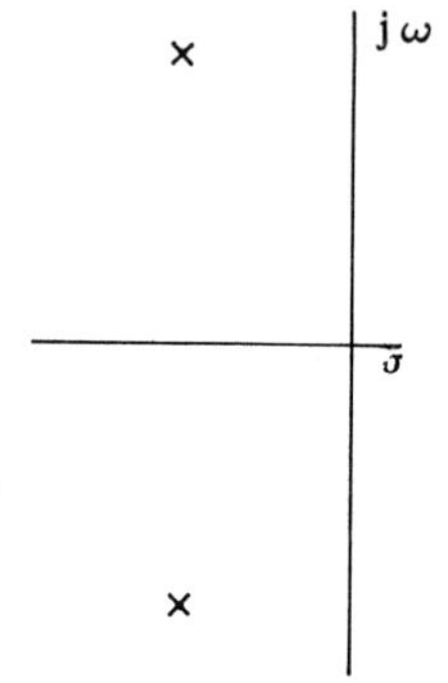

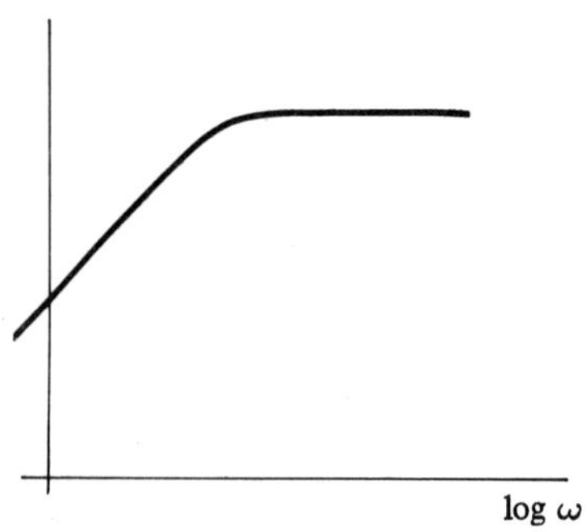

$$T(s) = \frac{s^2}{s^2 + \dfrac{\omega_p}{Q_p}s + \omega_p{}^2}$$

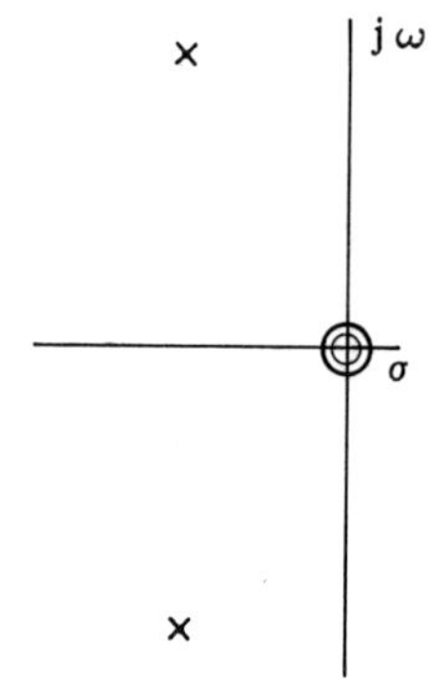

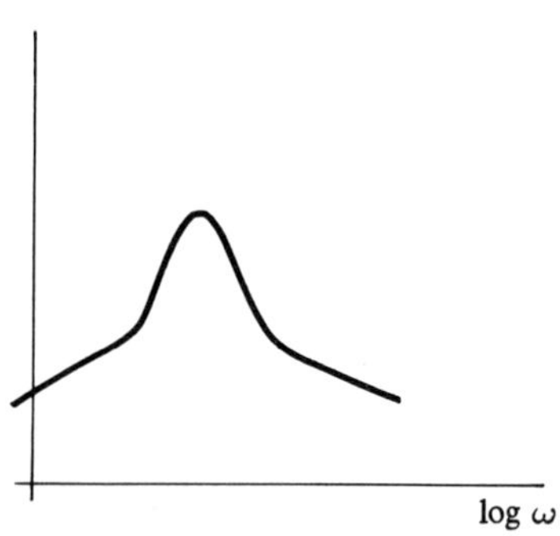

$$T(s) = \frac{s}{s^2 + \dfrac{\omega_p}{Q_p}s + \omega_p{}^2}$$

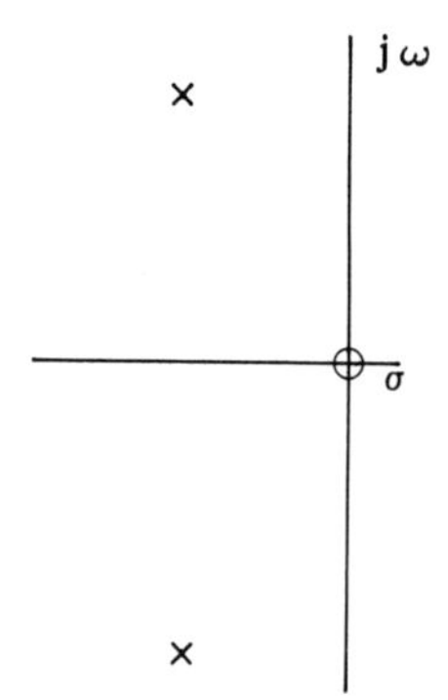

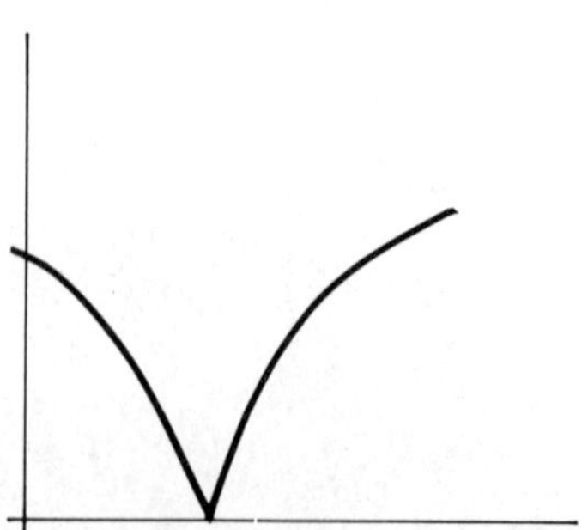

$$T(s) = \frac{(s^2 + \omega_O)^2}{s^2 + \dfrac{\omega_p}{Q_p}s + \omega_p{}^2}$$

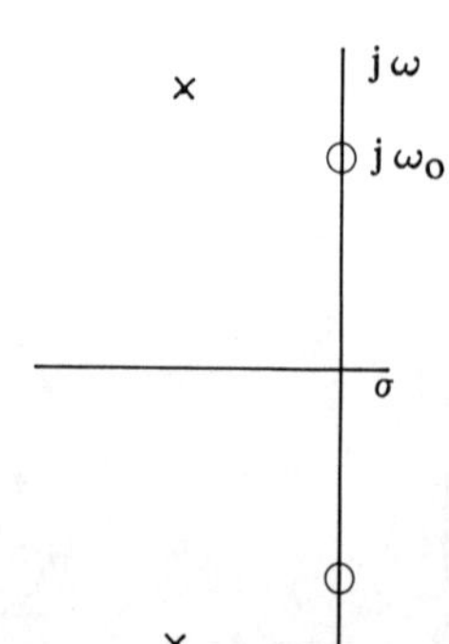

(g)

Fig. 4.22 Bode plots and pole zero diagrams for systems having complex conjugate pairs of poles.

The example shown in Fig. 4.22b, is often found in second-order systems having two time constants and feedback. There are only the two poles with the resulting constant gain at low frequencies and 40 dB/decade fall at high frequencies; the response at frequencies near to $\omega = a$ will again depend upon the Q factor.

The third example, (c), has a real negative pole as well as the complex conjugate pair which have a high Q factor and a resonant frequency greater than the real pole frequency. The low frequency gain is constant and the high frequency fall is 60 dB/decade.

Diagram (d) shows a pair of complex zeros and a triple pole at a higher frequency. As there is neither a pole nor a zero at the origin, the low frequency gain is constant. The complex zeros produce a resonant dip followed by a rise of 40 dB/decade up to the triple pole frequency, which results in a final high frequency fall at 20 dB/decade.

Response (e) shows the 'dog ears' obtained with an RC coupled amplifier with feedback. The low frequency pair of poles together with the zeros at the origin correspond with the low frequency resonant rise while the high frequency pair of complex poles produce the high frequency resonant rise and the following 40 dB/decade fall.

Filter circuits are synthesized by combining one or more pairs of complex poles with or without a real pole. For a high pass filter, zeros will be included at the origin. Example (f) shows the pole pattern for a low pass filter. The five poles produce a roll-off of 100 dB/decade. The resonant rise of the high Q pair compensate for the fall-off due to the real pole.

Example (g) compares the pole zero patterns for the four basic filter types. These are low pass, high pass, band pass and band stop respectively. The examples shown are second order (two pole) and the transfer functions are expressed in terms of the cut-off frequencies and Q factors. When a sharper or more selective cut-off is required, the transfer functions will be of a higher order with the pole zero pattern determined by the particular filter specification that is required.

4.15 SUMMARY OF THE EFFECTS OF THE POLE ZERO DIAGRAM ON THE STEADY-STATE RESPONSE

Low Frequency Response (range below lowest finite poles and zeros).

i. No poles or zeros at the origin; gain is constant at $T(0)$ obtained by substituting $s = 0$ in $T(s)$; no phase change.

ii. N_z zeros at the origin; d.c. gain is zero and gain rises at $N_z \times 20$ dB/decade (or $N_z \times 6$ dB/octave); low frequency phase change is $N_z \times (+90°)$.

iii. N_p poles at the origin; d.c. gain is infinite and gain falls at $N_p \times 20$ dB/decade (or $N_p \times 6$ dB/octave); low frequency phase change is $N_p \times (-90°)$.

Medium Frequency Response (the range in which finite poles and zeros lie).

i. Slope of gain response with increase in frequency changes by +20 dB/decade (or +6 dB/octave) at each real zero and by –20 dB/decade (or –6 dB/octave) at each real pole frequency; phase change tends towards an additional +90° for each real zero and an additional –90° for each real pole.

ii. Pairs of complex conjugate poles or zeros result in the same overall effects as in (i) above but additionally produce a resonant rise or fall if the angle θ made with the negative real axis is greater than 60°; if this angle is less than 60°, their effect is approximately the same as that for a pair of real poles or zeros.

High Frequency Response (range above highest finite pole or zero).

i. Final slope is –20 dB/decade × $(N_p - N_z)$ (or –6 dB/octave × $(N_p - N_z)$) where N_p and N_z are the total number of poles and zeros respectively.
ii. Final phase change is $-90° \times (N_p - N_z)$.

Spot Frequency Response. The response at single frequencies can be obtained from:

$$T(j\omega) = K \frac{\left[\text{The product of the directed line segments from each zero to the specified } j\omega.\right]}{\left[\text{The product of the directed line segments from each pole to the specified } j\omega.\right]}$$

4.16 EXAMPLES FOR FURTHER PRACTICE

Example 4.8. For each of the following transfer functions, determine $T(j\omega)$ at the specified values.

a) $\dfrac{5\,(s+2)}{(s+4)}$, at $\omega = 0$, 2 and 10 rads^{-1}.

b) $\dfrac{10^{13}s^2}{(s+200)^2\,(s+10^5)\,(s+2\times 10^5)}$, at $\omega = 0$, 200, 5000 and 10^5 rads^{-1}.

c) $\dfrac{20\,(s+1.5)}{s\,(s+0.1)\,(s+4)\,(s+20)}$, at $\omega = 0$, 0.5 and 5 rads^{-1}.

(2.5, 3.16$\angle$18.4°, 4.73$\angle$10.5°; 0, 250$\angle$90°, 499$\angle$0.3°, 316$\angle$–71.6°; ∞, 1.54$\angle$–159°, 0.0317$\angle$–171°.)

Example 4.9. For the following two transfer functions, calculate $|T(j\omega)|$ for $\omega = 50$, 95, 100 and 200 rads^{-1}.

a) $\dfrac{200s}{(s^2+20s+10^4)}$.

b) $\dfrac{50\,(s^2+50s+5000)}{(s+30)\,(s+100)}$.

(1.3, 8.9, 10, 2.4; 27.2, 22.6, 24, 40.3.)

Example 4.10. For each of the transfer functions in Examples 4.8 and 4.9 above, draw the pole zero diagrams and use the directed line segment method to confirm the answers obtained by substitution and calculation.

Example 4.11. For each of the following transfer functions, construct the Bode gain and phase plots. (The answers given for each case are, if applicable, the maximum gain (dB), the unity gain frequency ($\mathrm{rad\,s}^{-1}$) and the frequency at which the phase shift is $-180°$.)

a) $$\frac{5000}{(s+4)(s+6)(s+20)}.$$

b) $$\frac{6\times 10^{12}s^2}{(s+20)(s+100)(s+10^5)(s+3\times 10^5)}.$$

c) $$\frac{8\times 10^7(s+20)}{(s+110)(s+800)^2}.$$

d) $$\frac{4\times 10^8(s+20)^2}{s(s^2+110s+1000)(s^2+1300s+4\times 10^4)}.$$

(20.35, 13.5, 14.9; 46, 3.2 and 2.5×10^6, ∞; 40.4, 8900; ∞, 550, 700.)

Example 4.12. The following transfer functions include at least one pair of complex conjugate poles or zeros. Construct the Bode gain plot for each of these transfer functions. (The answers given for each case are, the gain maxima or minima and the frequency at which they occur together with the unity gain frequency.)

a) $$\frac{1000s}{(s^2+20s+500)}.$$

b) $$\frac{10^7}{(s^2+140s+10^6)}.$$

c) $$\frac{10^{10}}{(s+250)(s^2+300s+4\times 10^6)}.$$

d) $$\frac{10^7(s^2+200s+25\times 10^4)}{(s+2000)^2(s+10^4)}.$$

e) $$\frac{6.25\times 10^{15}}{(s^2+9397s+2.5\times 10^7)(s^2+3420s+2.5\times 10^7)}.$$

(34, 22.36, 0.5 and 1000; 37.1, 995, 3300; 18.45, 1975, 9.96, 1150, 2750; 27.3, 480, 57.7, 5600, 10^7; 20.13, 3200, 8900.)

Example 4.13. For each of the following transfer functions, draw the pole zero diagram and state, without calculation, the slope of the low and high frequency asymptotes, the low and high frequency angle and any other important features of the frequency response. Illustrate each with a sketch.

a) $$\frac{100}{(s+2)(s+4)(s+25)}.$$

b) $$\frac{10^{17}s^2}{(s+100)(s+1000)(s+10^5)^2}.$$

c) $$\frac{10^5\,(s+30)}{s\,(s+200)(s+500)}.$$

d) $$\frac{(s+4000)(s+20\,000)}{(s+250)(s+10^5)}.$$

e) $$\frac{10^{20}}{s\,(s+5000)^2\,(s+10^5)^2}.$$

f) $$\frac{s\,(s+40)\,10^7}{(s+10)(s+400)(s+2000)}.$$

g) $$\frac{700s}{(s^2+14s+700)}.$$

h) $$\frac{10^{10}}{(s^2+600s+8\times 10^5)(s+5000)}.$$

i) $$\frac{2\times 10^9}{(s^2+50s+8\times 10^5)(s+200)}.$$

j) $$\frac{6\times 10^{14}s^2}{(s^2+4s+100)(s^2+8000s+10^{10})}.$$

k) $$\frac{s^3}{(s+12\,000)(s^2+12\,000s+1.44\times 10^8)}.$$

(0, –60; 0, –270; constant l.f.: +40, –40; +180, –180; constant mid-band: –20, –40; –90, –180; constant mid-band: 0, 0; 0, 0; lag-lead mid-band: –20, –100; –90, –450: +20, –20; +90, –90: +20, –20; +90, –90; resonant rise at 26.5, Q 1.89: 0, –60; 0, –270; resonant rise at 894, Q 1.5: 0, –60; 0, –270; resonant rise at 894 Q 17.9: +40, –40; +180, –180, resonant rise at 10, Q 2.5 and at 10^5, Q 12.5: +60, 0; +270, 0; high pass filter.)

5

Root Locus Techniques

5.1 INTRODUCTION

In this chapter, the use of root locus methods is introduced with reference to the ideas gained in the preceding three chapters. The root locus is demonstrated for a simple system and as a consequence of the limitations found, a general 'root locus method' is introduced for negative feedback systems. A major part of the chapter is then devoted to a detailed study of the techniques of root locus plotting. A final section is then included to show how the techniques must be modified for positive feedback systems.

5.2 THE ROOT LOCUS PLOT

In Chapter 3, the transient response was shown to be determined by the poles and zeros of the system transfer function, together with the applied signal. In Chapter 4, the steady-state frequency response was also shown to be related to the poles and zeros of the system transfer function. In Chapter 2, the positions of the poles and zeros of a system transfer function were shown to depend upon the system components and other parameters such as the gain of amplifiers. Thus, if any parameter or component is changed, we can expect the system poles and zeros to move, or change their position on the s plane. If such changes can be predicted, the resulting effects upon the transient and steady-state responses can also be estimated. Alternatively, if a particular transient or steady-state response is required, the optimum value of a parameter or component can be selected.

A root locus plot is a locus on the s *plane for all the poles (or all the zeros) of the system transfer function, obtained by varying a particular parameter or component between zero and infinity.*

For simple systems the poles and zeros of the transfer function can be expressed as a function of the component of interest, and then by substituting values the locus can be constructed. This may be observed by means of an example.

Example 5.1. Determine the voltage transfer function for the circuit shown in Fig. 5.1 and hence construct the loci of the poles and zeros for this transfer function if R is varied between 0 and ∞.

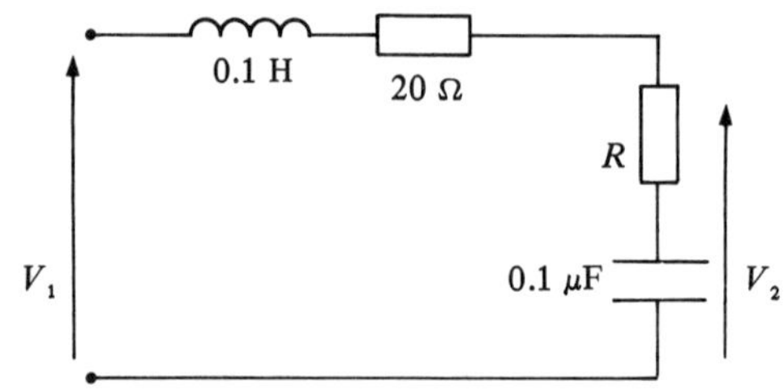

Fig. 5.1 A series *RLC* circuit for demonstration of the locus of transfer function poles in Example 5.1.

Solution. The transfer function can be shown to be given by:

$$T(s) = \frac{10R\,(s + 10^7/R)}{s^2 + (200 + 10R)\,s + 10^8} \,. \qquad \ldots(5.1)$$

The pole positions may be obtained by equating the denominator to 0 and solving the quadratic equation. The resulting position of the two poles is given by:

$$p_n = -\frac{(200 + 10R)}{2} \pm \sqrt{\left(\frac{200 + 10R}{2}\right)^2 - 10^8} \,. \qquad \ldots(5.2)$$

The position of the zeros can be obtained by equating the numerator to 0. In this case, there is only one zero and the position is given by:

$$z_1 = -10^7/R \,. \qquad \ldots(5.3)$$

The positions of both poles and the zero depend upon the value of R. We can therefore plot both a zero locus and a pole locus. (It is far more usual to find the zero of a transfer function independent of the parameter being varied, in which case there is only a root locus for the poles.)

From expression (5.2), when $R = 0$,

$$p_n = -100 \pm \mathrm{j}10^4 \,, \quad \text{and} \quad z_1 = -\infty \,.$$

Substitution of other values of R leads to the following results for the poles p_n and the zero z_1.

R (ohms)	p_n	z_1
500	−2600 ± j9656	−20 000
1500	−7600 ± j6500	−6667
1800	−9100 ± j4146	−5555
1980	−10 000, −10 000	−5050
2000	−8682, −11 517	−5000
5000	−2078, −48 121	−2000
∞	0 , −∞	0

These results are shown plotted in Fig. 5.2; Fig. 5.2a is the pole locus and Fig. 5.2b shows the locus of the zero. (Note equal scales for σ and jω axes.)

If R is set at 500 Ω and C is varied, the zero locus will have the same form, but the pole locus is somewhat different. It is left to the reader to calculate some values and find the new form. Yet another form will be obtained if R and C are fixed and the inductor value is chosen as the *parameter* for the loci.

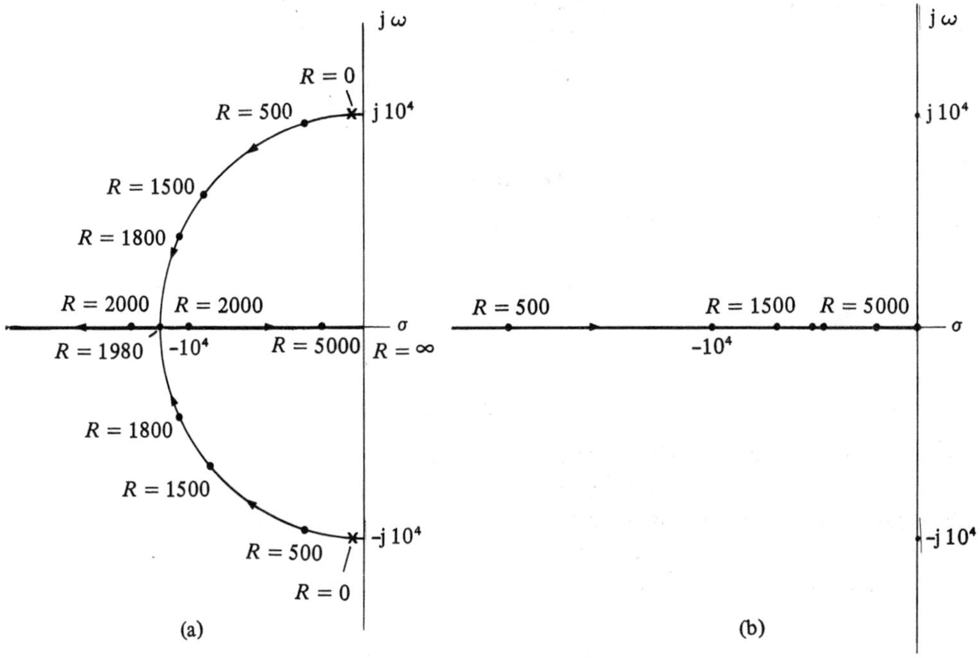

Fig. 5.2 Root locus plots for Example 5.1. a) Locus of the poles when R is varied between 0 and ∞. b) Locus of the zero when R is varied between 0 and ∞.

The plotting of the loci in the above example involved no more than the repeated solution of a quadratic equation. In many systems, the characteristic equation obtained by equating the denominator to zero will be a cubic or higher-order equation. Determining the roots or pole positions algebraically is no longer a simple matter. The root locus method avoids the solution of higher order equations and allows us to plot a locus with an acceptable level of accuracy by the application of a set of rules. This method can, however, be used to obtain the roots of a higher-order equation if necessary. This technique is discussed in the next chapter.

5.3 THE ROOT LOCUS FORM OF THE CHARACTERISTIC EQUATION

In general, the characteristic equation is given by a polynomial in s equated to zero. For example:

$$as^4 + bs^3 + cs^2 + ds + e = 0 \ . \qquad \ldots(5.4)$$

One or more of the coefficients in this expression will contain the parameter which is to be varied. The first step is to separate such terms from those not containing the parameter.

From (5.4), suppose the coefficients have the following values: $a = 1, b = 4, c = (2 + K)$, $d = 10K$ and $e = 4K$, where K is the parameter to be investigated. The *CE* becomes:

$$s^4 + 4s^3 + 2s^2 + K(s^2 + 10s + 4) = 0 \ . \qquad \ldots(5.5)$$

This form is often written,

$$q(s) + Kp(s) = 0 \ . \qquad \ldots(5.6)$$

The next step is to divide through by $q(s)$ to give:

$$1 + K\frac{p(s)}{q(s)} = 0 \ , \qquad \ldots(5.7)$$

or

$$1 + KG(s) = 0 \ . \qquad \ldots(5.8)$$

Note that in this context, $G(s)$ is the particular transfer function that has been obtained from the characteristic equation in this way. $G(s)$ is also used for the forward gain of a feedback control system. This is not inconsistent as for a unity feedback system, this is indeed the required form for expression (5.8).

Returning to our numerical example, expression (5.5) is rearranged to:

$$1 + \frac{K(s^2 + 10s + 4)}{s^2(s^2 + 4s + 2)} = 0 \ . \qquad \ldots(5.9)$$

5.4 THE BASIS FOR ROOT LOCUS CONSTRUCTION

Referring to expression (5.8),

$$KG(s) = -1 = 1\angle 180^\circ \ . \qquad \ldots(5.10)$$

This result forms the whole basis for the root locus method. $G(s)$ is simply a particular transfer function associated with the system (as mentioned above; in feedback systems, it is often the system open-loop transfer function). The angle and modulus of any transfer function depend upon the value of s and can be found from the pole zero diagram of $G(s)$. (See results (4.28) and (4.29) in the last chapter.)

Thus, if $\angle G(s)$ is found for many values of s and if in some cases $\angle G(s) = 180°$, the values of s in these cases must be points on the locus. This process can be referred to as *the angle test.* In the most general approach, the angle test is applied to a very large number of points in the s plane and those points which satisfy the test form the required locus. This obviously would be a most tedious procedure (although it can be the basis for computer programmed plotting of loci) and there are many simplifying rules introduced later in this chapter.

Having found the position of the locus by means of the angle test, we also require the values of the parameter K which correspond to particular points on the locus. For any particular point, from expression (5.10):

$$|KG(s)| = 1 \,, \qquad \ldots(5.11)$$

and as K will be a real number;

$$K = \frac{1}{|G(s)|} \,. \qquad \ldots(5.12)$$

This result can also be obtained from the pole zero diagram for $G(s)$ using result (4.31). In this case, however, as the reciprocal of $G(s)$ is required;

$$K = \frac{\left[\begin{array}{l}\text{Product of the lengths of the directed line segments}\\ \text{from each } \textit{pole} \text{ to the particular point on the locus}\end{array}\right]}{\left[\begin{array}{l}\text{Product of the lengths of the directed line segments}\\ \text{from each } \textit{zero} \text{ to the particular point on the locus}\end{array}\right]} \,. \qquad \ldots(5.13)$$

This is known as *the magnitude test.*

5.5 FURTHER PROPERTIES OF THE ROOT LOCUS

i. Since the original characteristic equation to be investigated is nth order (expression (5.4)), there will be n roots for any value of K. *There will therefore be n branches of the locus* each following the change of a particular root as K is changed. ...(5.14)

ii. Referring to expressions (5.5) and (5.6), if K is zero, the locus starts from the roots of $q(s)$ which are the poles of $G(s)$ (5.7 and 5.8). As K tends to infinity, $q(s)$ becomes negligible (5.6): the locus then finishes at the roots of $p(s)$, which are the zeros of $G(s)$ (5.7 and 5.8). *As K is increased from zero to infinity, the locus moves from the poles of G(s) to the zeros of G(s).* Note, this may include poles or zeros at infinity. This will be discussed in v. below. ...(5.15)

iii. The polynomials cannot have the same root for two different values of K. Thus a branch of a locus cannot cross itself to form a closed loop. Two or more equal roots are possible, however, thus two or more branches of a locus may meet and then separate. This is usually two real axis branches coming together and then separating to become complex.

iv. Since any complex roots must be in conjugate pairs, any complex part of a locus must have a conjugate part. This means that *the locus must be symmetrical about the real axis.* ...(5.16)

v. If $G(s)$ has more finite poles than finite zeros, there will be sufficient zeros at infinity to make the total number of each equal (Chapter 2, p. 41). The angle with respect to the real axis, at which these zeros lie depends upon their number; as the locus approaches these zeros, the angle test must still apply. This requires that *the angles of the asymptotes to infinity add up to 180°* or, more generally: if there are N_p poles and N_z zeros of $G(s)$, there will be $(N_p - N_z)$ asymptotes to infinity at angles with respect to the real axis of:

$$\frac{(1 \pm 2n)\,180^\circ}{N_p - N_z} . \qquad \ldots(5.17)$$

In practice, there are rarely more than five asymptotes and the angles for up to five are given in the following table.

$N_p - N_z$	*Angles of asymptotes*
1	180°
2	± 90°
3	180°, ± 60°
4	± 45°, ± 135°
5	± 36°, ± 108°, 180°

vi. It may also be shown that these asymptotes meet on the real axis at:

$$s = \frac{\text{Sum of the poles of } G(s) - \text{Sum of the zeros of } G(s)}{\text{Number of asymptotes}} . \qquad \ldots(5.18)$$

Result (5.18) includes any complex poles or zeros of $G(s)$ but only the real parts are of importance as the imaginary parts will always cancel.

If N_z is greater than N_p, there will be $N_z - N_p$ poles at infinity. The rules for the angles and meeting points of the asymptotes *from* infinity are the same as those for the asymptotes to infinity as described in rules (5.17) and (5.18) above.

vii. If the angle test is applied to points on the real axis, the angles from any complex poles and zeros will always cancel and any real poles or zeros to the left of the point will contribute 0°. Each pole or zero to the right will contribute 180°. Thus, *the parts of the real axis to the left of an odd number of poles and zeros will be a part of the locus.* ...(5.19)

5.6 SKETCHING THE APPROXIMATE LOCUS

Rules (5.14) to (5.19) are usually sufficient to obtain a sketch of the general form of the locus. This should always be done as it will be of considerable help in deciding how best to fill in more detail. This stage in the procedure can now be illustrated with some examples.

Example 5.2. A number of functions have been arranged in the root locus form with the values of $G(s)$ listed below. In each case, sketch the general form of the locus.

a) $$G_1(s) = \frac{(s+1)(s+2)}{s(s+3)(s+4)} .$$

b) $$G_2(s) = \frac{1}{s(s+1000)} .$$

c) $$G_3(s) = \frac{s}{(s+20)(s+35)(s+40)} .$$

d) $$G_4(s) = \frac{(s+8)}{s(s+2)(s+3)} .$$

e) $$G_5(s) = \frac{(s+1)(s+2)}{s(s+0.4)} .$$

f) $$G_6(s) = \frac{1}{s(s+2000)(s+3000)} .$$

g) $$G_7(s) = \frac{1}{(s+3)(s+4)(s^2+2s+5)} .$$

h) $$G_8(s) = \frac{(s+0.1)}{s(s+0.01)(s+1)(s+5)} .$$

i) $$G_9(s) = \frac{s^3}{(s+10)(s+30)^2} .$$

Solution. For each of these examples, the first steps are as follows:

i. Plot the poles and zeros of $G(s)$ on the s plane – starting and finishing points of the locus from (5.15).

ii. Mark the parts of the real axis to the left of an odd number of poles and zeros as part of the locus (5.19).

iii. Determine the number and position of asymptotes to infinity or from infinity (5.17 and 5.18).

a) $$G_1(s) = \frac{(s+1)(s+2)}{s(s+3)(s+4)} .$$

The zeros of $G_1(s)$ are at -1 and -2 and the poles are at 0, -3 and -4. These are plotted in Fig. 5.3a. The negative real axis between 0 and -1 is to the left of *one* pole and it is therefore a part of the locus. Between -1 and -2 it is to the left of a pole and a zero, an even number and is therefore not part of the locus. Similarly, between -2 and -3 is to the left of three 'poles and zeros', and between -4 and infinity is to the left of five; these parts are also parts of the locus. Between -3 and -4 is to the left of an even number and is therefore not part of the locus.

There are two zeros and three poles; there must be one asymptote to infinity lying at 180°. As this is along the real negative axis, the meeting point is irrelevant. In any case, we have already found that the locus runs along this axis between -4 and infinity. We can see that there are three branches of the locus starting from the poles of $G(s)$ and going to the zeros of $G(s)$; thus, the locus is complete.

b)
$$G_2(s) = \frac{1}{s(s+1000)} .$$

Fig. 5.3b shows the poles of $G(s)$ (no finite zeros) and the parts of the locus on the real axis plotted as before. There are two asymptotes at $\pm 90°$, (5.17). The meeting point of the asymptotes on the real axis is given by:

$$s = \frac{0 - 1000}{2} = -500 .$$

The two branches of the locus start from 0 and -1000. They come together as K is increased until, at a particular value of K, they meet. Further increase in K results in a complex conjugate pair of branches approaching infinity along the asymptotes. *The point of departure from the real axis* will be considered later, but in this case, it is reasonable to suppose that it will occur at a point equidistant from zero and -1000. In fact, this point of departure will coincide with the asymptote meeting point and the two branches go straight along the asymptotes. This can be confirmed by consideration of the *CE* from which $G(s)$ came:

$$s(s+1000) + K = 0 ,$$

or

$$s^2 + 1000s + K = 0 .$$

The roots or poles lie at $-500 \pm \sqrt{25 \times 10^4 - K}$. Now, if K is greater than 25×10^4, the roots are complex with real parts that are always -500.

c)
$$G_3(s) = \frac{s}{(s+20)(s+35)(s+40)} .$$

The poles and zeros and real parts are plotted as before in Fig. 5.3c. There are again two asymptotes at $\pm 90°$ and they meet the real axis at $s = \dfrac{-20-35-40}{2} = -47.5$. There must be three branches of the locus: the first goes from the pole at -20 to the zero at the origin, the other two start from the poles at -35 and -40, come together and then separate to approach the asymptotes. The approach may be slow or fast; as yet, we have no means of telling. The locus shown in Fig. 5.3c is typical of this situation.

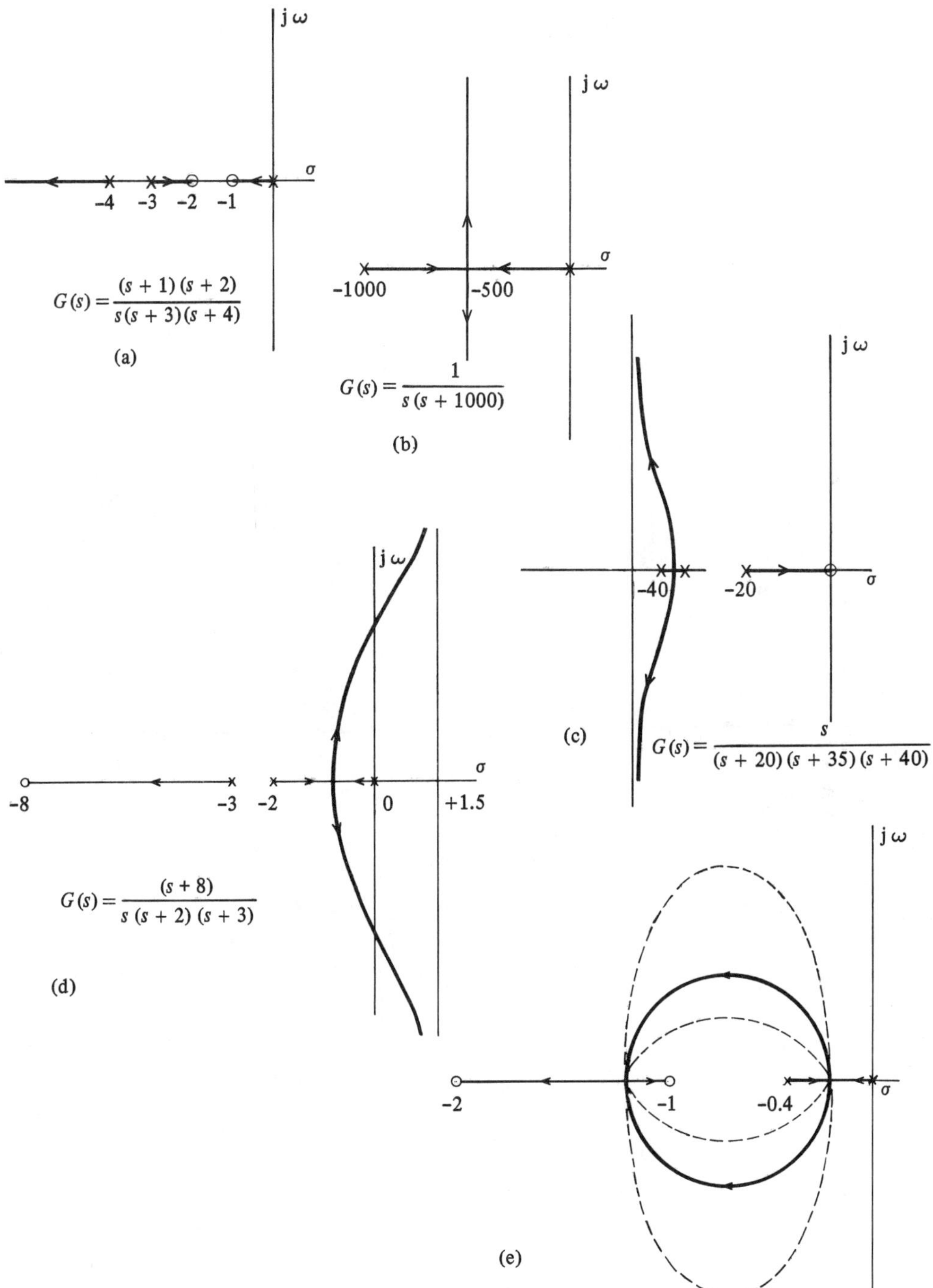

j ω
σ
-4 -3 -2 -1
$G(s) = \frac{(s+1)(s+2)}{s(s+3)(s+4)}$
(a)
j ω
σ
-1000
-500
$G(s) = \frac{1}{s(s+1000)}$
(b)
j ω
-40
-20
σ
(c)
$G(s) = \frac{s}{(s+20)(s+35)(s+40)}$
j ω
σ
-8
-3 -2
0
+1.5
$G(s) = \frac{(s+8)}{s(s+2)(s+3)}$
(d)
j ω
-2
-1
-0.4
σ
(e)

Fig. 5.3

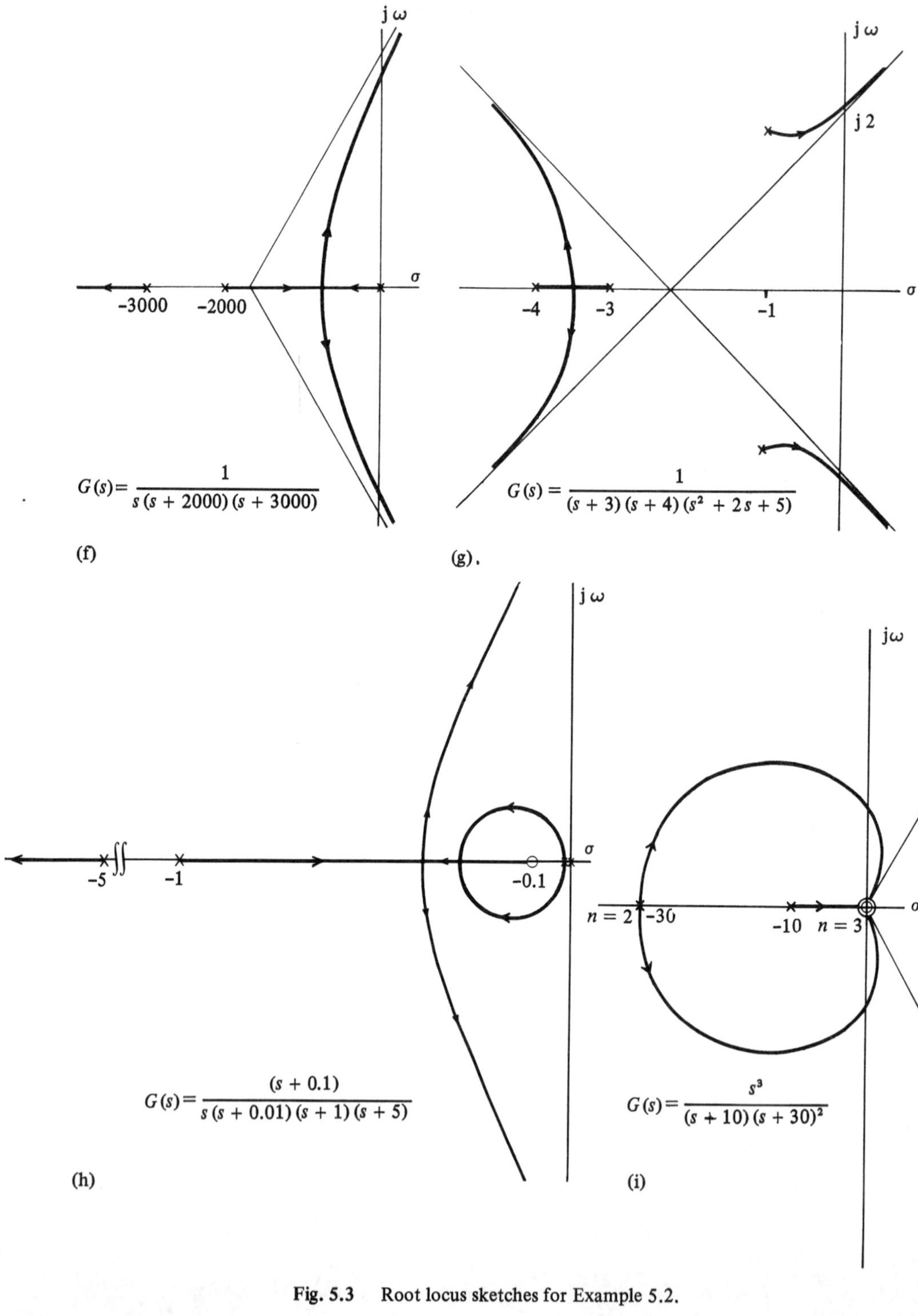

Fig. 5.3 Root locus sketches for Example 5.2.

d)
$$G_4(s) = \frac{(s+8)}{s(s+2)(s+3)} \,.$$

This is similar to the last case as there are three poles and one zero. The two asymptotes now meet at $s = \dfrac{0-2-3-(-8)}{2} = +1.5$. Once again, for an accurate plot, we should need to know the point of departure from the real axis. It would also be useful to know the $j\omega$ *axis crossing points* for the complex parts of the two branches of the locus since the value of K, beyond which the system becomes *unstable*, can be determined. The approximate form for the locus is shown in Fig. 5.3d.

e)
$$G_5(s) = \frac{(s+1)(s+2)}{s(s+0.4)} \,.$$

As $G_5(s)$ has an equal number of poles and zeros, there are no asymptotes to infinity. However, if rule (5.15) regarding the starting and finishing points is to be satisfied, the two branches of the locus must be complex for some values of K. The branches must leave the real axis at a point between the two poles, i.e. between 0 and –0.4 and then return to the real axis at a point between the zeros at –1 and –2. Thus, *the point of arrival on the real axis* must be investigated. We also have no means of predicting the complex conjugate paths between the points of departure and arrival. Fig. 5.3e shows several possibilities which apparently satisfy rules (5.14)–(5.19).

f)
$$G_6(s) = \frac{1}{s(s+2000)(s+3000)} \,.$$

This transfer function has three poles and no zeros. The three asymptotes lie at $\pm 60°$ and $180°$, meeting at $s = -1667$ as shown in Fig. 5.3f. Once again, a more accurate plot will require a point of departure from the real axis and crossing points for the imaginary axis.

g)
$$G_7(s) = \frac{1}{(s+3)(s+4)(s^2+2s+5)} \,.$$

There are four poles and no zeros in this case. The four asymptotes lie at $\pm 45°$ and $\pm 135°$. The four poles are at –3, –4, $-1+j2$ and $-1-j2$. The meeting point for the asymptotes is:

$$s = \frac{-3-4-1+j2-1-j2}{4} = -2.25 \,.$$

The asymptotes and poles are shown in Fig. 5.3g. The branches of the locus starting from –3 and –4 will clearly approach the $\pm 135°$ asymptotes as shown. Determination of the other two branches will require the j axis crossing point and *the angle of departure from a complex pole,* which is discussed later in this chapter.

h)
$$G_8(s) = \frac{(s+0.1)}{s(s+0.01)(s+1)(s+5)} \,.$$

Plotting of the real parts and asymptotes here, initially suggests a simple form of locus, with the branches near the origin joining and dividing with complex paths nearly parallel to the imaginary axis. The alternative is shown in Fig. 5.3h; the two complex branches return to the real axis and divide again. One branch goes to the finite zero at –0.1 and the other meets the branch coming from the pole at $s = -1$ and then redivides to approach the asymptotes. This form is, in fact, the correct one which can be confirmed later when other rules have been demonstrated.

i) $$G_9(s) = \frac{s^3}{(s+10)(s+30)^2} \,.$$

For this final example, there are no asymptotes and the locus lies on the real axis between the origin and $s = -10$. The other two branches must leave the poles at $s = -30$ (at right angles?) and take conjugate paths to two of the three zeros at the origin. Common sense suggests that they should approach the origin at ±60° as shown. This will be confirmed later.

5.7 MORE ACCURATE LOCUS PLOTTING

In the above examples, a number of areas for investigation are suggested. These are as follows:

i. Crossing points on the imaginary axis.
ii. Points of arrival on and departure from the real axis.
iii. Angles of arrival and departure from a single point on the real axis (not illustrated in the above examples).
iv. Angles of arrival or departure from coincident real zeros or poles.
v. Angle of departure from a complex pole or arrival at a complex zero.
vi. Confirmation of any part of the locus.

We shall now take these points in turn and with the help of examples deduce further rules or procedures so that an accurate plot can be constructed.

i. *Determination of crossing points on the imaginary axis.* This is simply achieved by application of the angle test (5.10) to points along the imaginary axis. A reasonable guess can often be made by examination of the asymptotes. An acceptable point can then be found after two or three calculations (or measurements).* Fig. 5.4 illustrates this for $G(s) = \dfrac{1}{(s+1)(s+3)(s+6)}$. The asymptotes and real parts are shown.

Try a point $s_1 = j4$. $\angle s_1 = -\tan^{-1}4 - \tan^{-1}4/3 - \tan^{-1}4/6 = -163°$ (angles shown).

Try $s_2 = j5.5$, $\angle s_2 = -184°$.

Try $s_3 = j5.0$, $\angle s_3 = -177°$.

Finally at $s_4 = j5.2$, $\angle s_4 = -180°$.

For most purposes, an angle within 1° of 180° is sufficiently accurate.

* See note at end of preface.

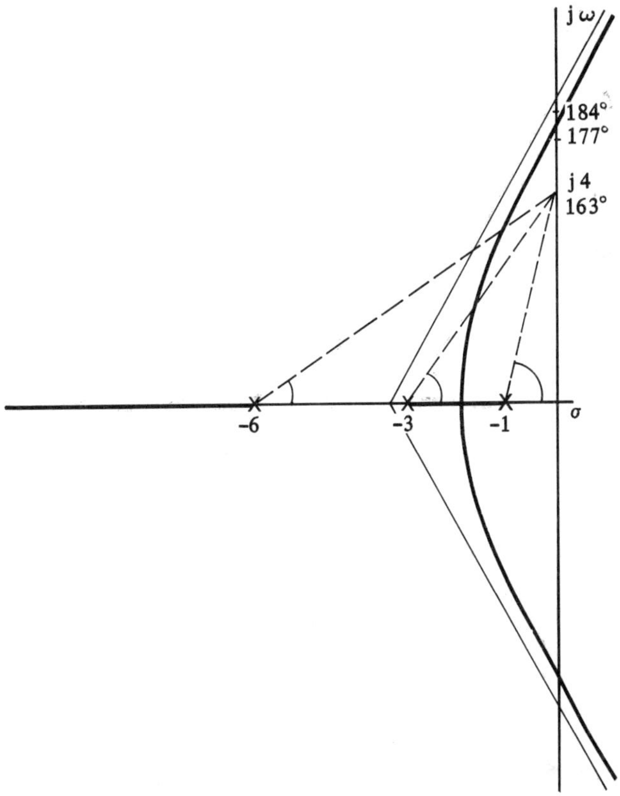

Fig. 5.4 Application of the angle test to determine the imaginary axis crossing point.

ii. *Points of arrival and departure on the real axis.* These are determined by rule (5.15) and by the magnitude test (5.13). Consider a part of the real axis which is also a part of the locus between two real poles. As K is increased from zero, two branches of the locus move from the poles until, at some value of K, the two branches meet. Further increase in K causes the branches to separate and become complex. Thus, *the point of departure from the real axis corresponds to the point at which K is a maximum between two poles along the real axis.* . . . (5.20)

This can be illustrated by referring again to Fig. 5.4. Using rule (5.13), the magnitude test, for a number of points along the real axis between $s = -1$ and $s = -3$, the value of K corresponding to each point can be found. For example, at $s = -1.5$, $K = |\frac{1}{G(s)}| = 0.5 \times 1.5 \times 4.5 = 3.375$. Similarly, K can be calculated for other values of s with the following results:

s	K	*Guess*
−1.5	3.375	1
−1.8	4.032	4
−1.9	4.059	5
−2.0	4.0	3
−2.5	2.625	2

In this, from rule (5.20), it is obvious that the point of departure is at approximately $s = -1.9$. Note that the sequence of guesses attempted to 'bridge' the required maximum. After results of 3.375, 2.625 and 4.0, the required point will probably lie between 3.375 and 4.0, but nearer to 4.0. Finally $s = -1.8$ and -1.9 have been tried to give the acceptable accuracy.

A similar reasoning and a similar rule apply to the point of arrival of the locus between two real axis zeros. In this case, *the point of arrival at the real axis corresponds to the point at which K is a minimum between two zeros along the real axis.* ...(5.21)

The reason for this is that any smaller values of K result in complex conjugate poles and larger values produce two real roots nearer to the zeros.

In some cases (see Fig. 5.3h), the points of arrival and departure occur along the same section of the real axis part of the locus. These are between a pole and a zero and correspond to a minimum and a maximum value of K. This may be difficult to detect, in which case a graph of K against s can be helpful. This technique is illustrated in Fig. 5.5 where the locus for $G(s) = \dfrac{(s+4.8)}{s\,(s+5)^2}$ is constructed. The locus only lies along the real axis between the origin and -4.8. There are two asymptotes at $\pm 90°$, meeting the real axis at -2.6. Although the locus might go directly form the double pole to the asymptotes, this does not seem very likely as, if one of the poles at -5 was moved by a small amount, it would cancel the zero and for the resulting system, the locus would leave the real axis along the asymptotes. The alternative is that the branches from the double pole rejoin the real axis. They would then separate, one going to the zero and the other joining the branch from the origin to again divide near the asymptote meeting point. To check this alternative, a graph of K values for different s values has been drawn in Fig. 5.5b. For example, at $s = -1$,

$$K = \frac{1 \times 4 \times 4}{3.8} = 4.21 .$$

This graph shows both a maximum (at $s = -2.65$) and a minimum (at $s = -4.6$) which confirms, respectively, a point of departure and a point of arrival on the real axis.

An alternative method is to write an equation for K in terms of s, differentiate and equate to zero. As this result is likely to require the solution of a higher-order equation (a cubic in the above example), it is rarely worth the effort.

iii. *Angles of departure or arrival from a single point on the real axis.* In some cases, the points of arrival and departure coincide. This is indicated on the K graph by a point of inflection. This is illustrated by the drawing of part of a locus (not to scale) and the associated K graph in Fig. 5.6.

This raises another problem; if y branches meet at a point on the real axis and y branches leave from the same point, what is the angle between these branches at the meeting point?

It can be shown that the angle λ between y approaching branches is given by:

$$\lambda_y = \pm\frac{360°}{y} . \qquad \ldots(5.22)$$

Also, the angle θ between a branch leaving and an adjacent branch approaching is given by:

$$\theta_y = \pm\frac{180°}{y} . \qquad \ldots(5.23)$$

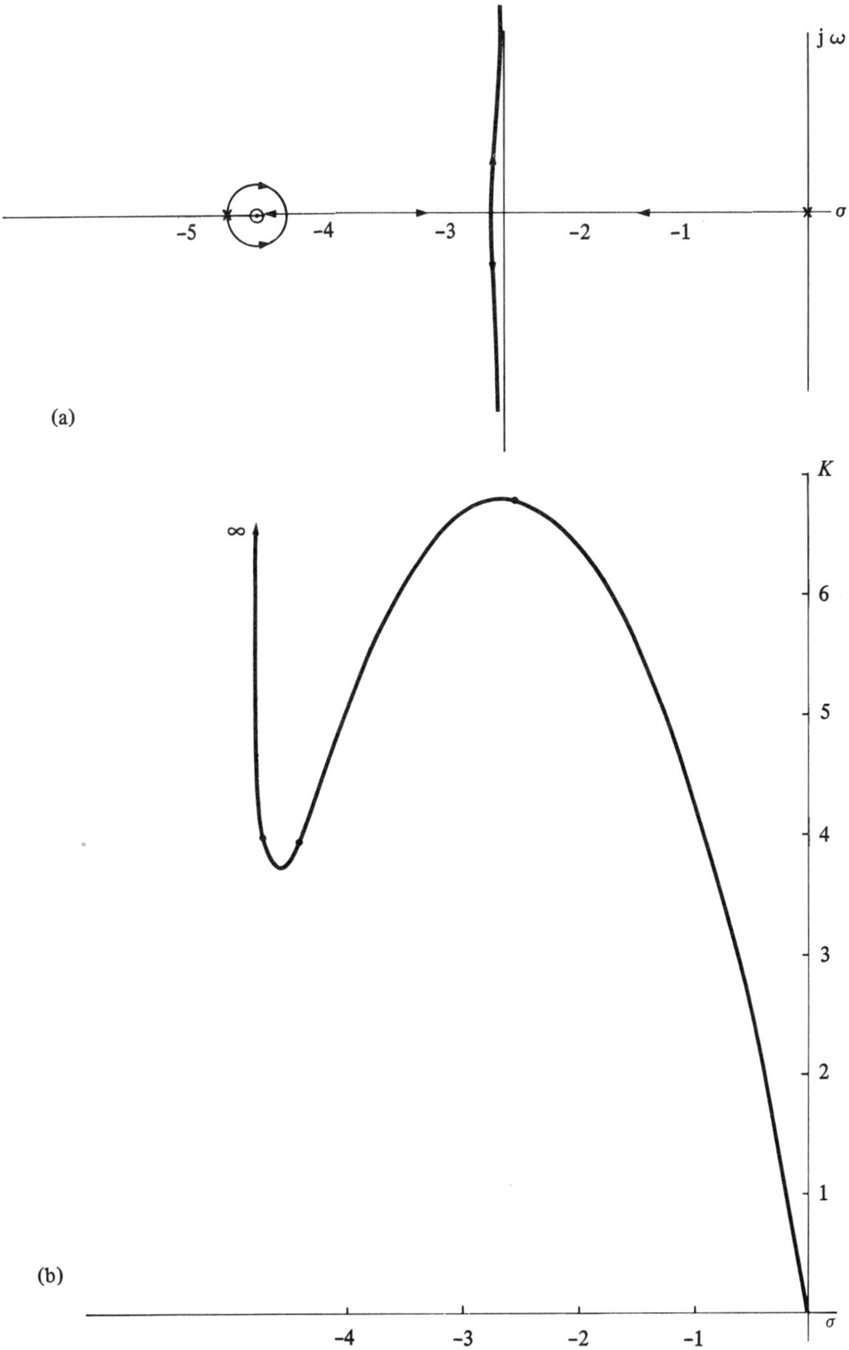

Fig. 5.5 Determination of the real axis points of arrival and departure by the hill and valley graphical method. a) The locus for $G(s) = \dfrac{(s+4.8)}{s\,(s+5)^2}$. b) The graph of K values for points along the real axis part of the locus.

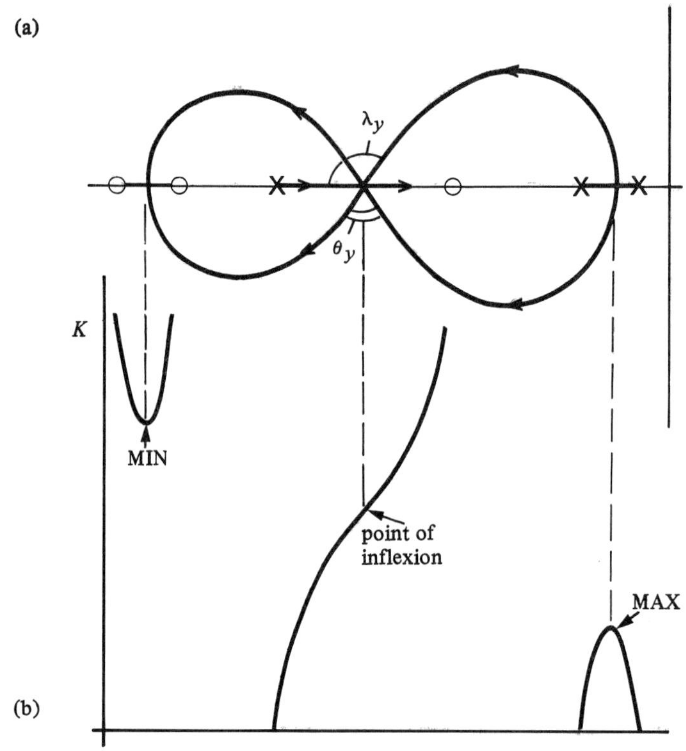

Fig. 5.6 The angle of departure from the real axis when the points of arrival and departure coincide. a) The locus. b) K values.

In the example shown in Fig. 5.6, there are three branches approaching the meeting point:

$$\therefore \qquad \lambda_y = \pm\frac{360^\circ}{3} = \pm 120^\circ \;.$$

One branch lies on the real axis coming from 180°; the other two must come from ±60°.

$$\theta_y = \pm\frac{180^\circ}{y} = \pm 60^\circ \;.$$

The leaving branches therefore have an angle of 60° with respect to the approaching branches. Another example of this can be seen with a $G(s)$ of $\dfrac{1}{(s+4)(s+8)(s^2+12s+40)}$. It is left to the reader to examine the locus resulting from this $G(s)$.

iv. *Angles of departure or arrival at coincident real poles or zeros.* Another similar situation can be observed in Example 5.2i above. We can see in Fig. 5.3i that three branches approach the triple zero at the origin (on the real axis). Rule (5.22) applies and the branches meet as shown, coming from 180° and from ±60°.

The last three sections have been concerned with branches of the locus meeting and separating at points along the real axis. These phenomena can also occur along complex branches of the locus. An example, where $G(s)$ is given by $\dfrac{1}{(s+1)(s+3)(s^2+4s+13)}$ is shown in Fig. 5.7. Methods similar to those described above can be applied, but as such a locus is relatively rare it need be discussed no further.

v. *Angle of departure from a complex pole or arrival at a complex zero.* This example was illustrated in Example 5.2g, but a better example is given by:

$$G(s) = \frac{(s+8)}{(s+1)(s^2+4s+29)}\ .$$

The poles of this $G(s)$ are shown in Fig. 5.8 together with the locus on the real axis and the two asymptotes at $\pm 90°$ which meet the real axis at $s = \dfrac{-1-2-2-(-8)}{2} = +1.5$.

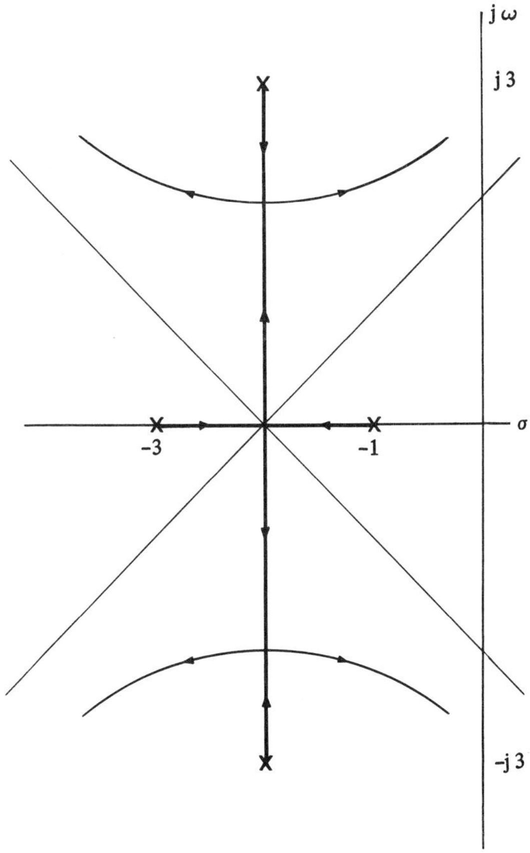

Fig. 5.7 An example of the locus breaking away from a complex branch.

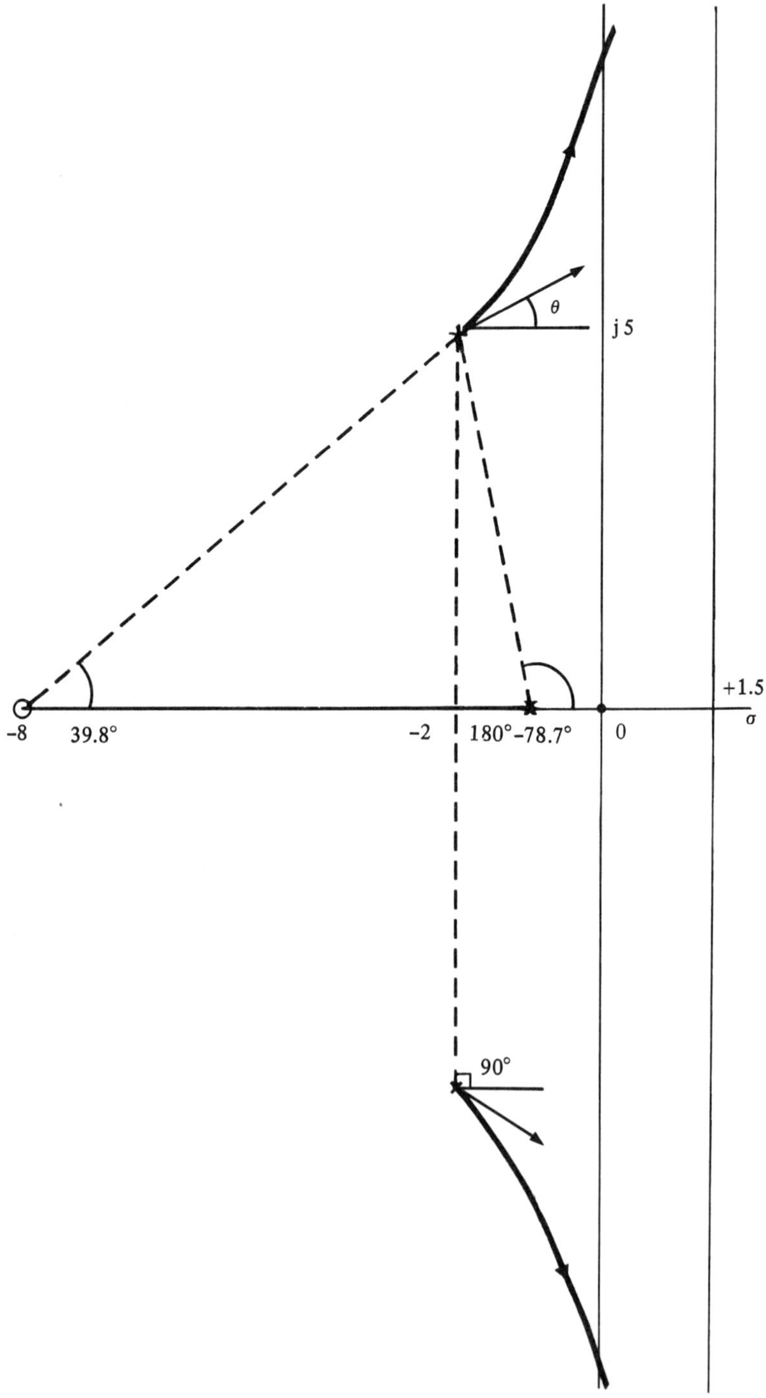

Fig. 5.8 Determination of the angle of departure from a complex pole.

The locus must leave the complex poles and cross the imaginary axis before approaching the asymptotes. It would be helpful if the angle of departure θ, from the complex pole could be found. A relatively quick and simple procedure can be used and as it does not involve 'trial and error' it should be applied before determination of the crossing point.

Consider a point s_1 very near to the pole at $s = -2 + j5$. This point is so near to this pole that, 'seen' from the other poles and zeros, it appears to coincide with that particular pole. If it is on the locus, it must lie at a particular angle θ with respect to the pole and it must satisfy the angle test (5.10). The angles from the other poles and zeros of $G(s)$ can be taken approximately to the pole very near to s_1. Thus, using the angles shown in Fig. 5.8,

[39.8°	– (180° – 78.7°)	–90°	–θ]	= 180° .	. . . (5.24)
from the zero	from a pole	from a pole	from the adjacent pole		

∴ the angle of departure θ, is given by:

$$\theta = 39.8° + 78.7° - 180° - 180° - 90° = 28.5°.$$

(taking 360° = 0°)

Although 28.5° is the *angle of departure,* the locus can move away from this straight line quite quickly as indicated in the diagram.

The angle of arrival at a complex zero is determined in a similar manner. The only difference is that θ will be an angle from a zero and will therefore be positive in an equation similar to (5.24) (the 90° will also be positive as this will be from a complex conjugate zero).

The reader should try this technique on Example 5.2g and confirm that the angle of departure is –349° or +11°.

vi. *Confirmation of detail on any part of the locus.* Any point can be checked by the angle test, but random selection of points would be very time-consuming and frustrating. When the general form, real axis meeting points, etc., have been determined, one or two more confirming points are usually sufficient. It is often helpful to draw a line which must cut the locus and then to confine the search to points along this line. An example is shown in Fig. 5.9 where the locus is being constructed for $G(s) = \dfrac{(s+5)(s+6)}{s(s+3)(s+7)}$. The parts of the real axis have been drawn and the arrival and departure points have been calculated. The path between these two must now be determined. A vertical line has been drawn through the point $s = -4$; then the angle test has been applied at points along this line with the results shown. The sequence was j3, j1, j1.5 and finally j1.3 where the angle was within 180 ± 1°. This is sufficient to sketch the remainder of the locus as shown.

5.8 SCALING THE LOCUS

Once the shape of the locus is known, some scaling is usually necessary. It may be sufficient to find the value of K above which the system will be unstable. Alternatively, a K value resulting in critical damping or even the value which produces the Q factor of unity may be required.

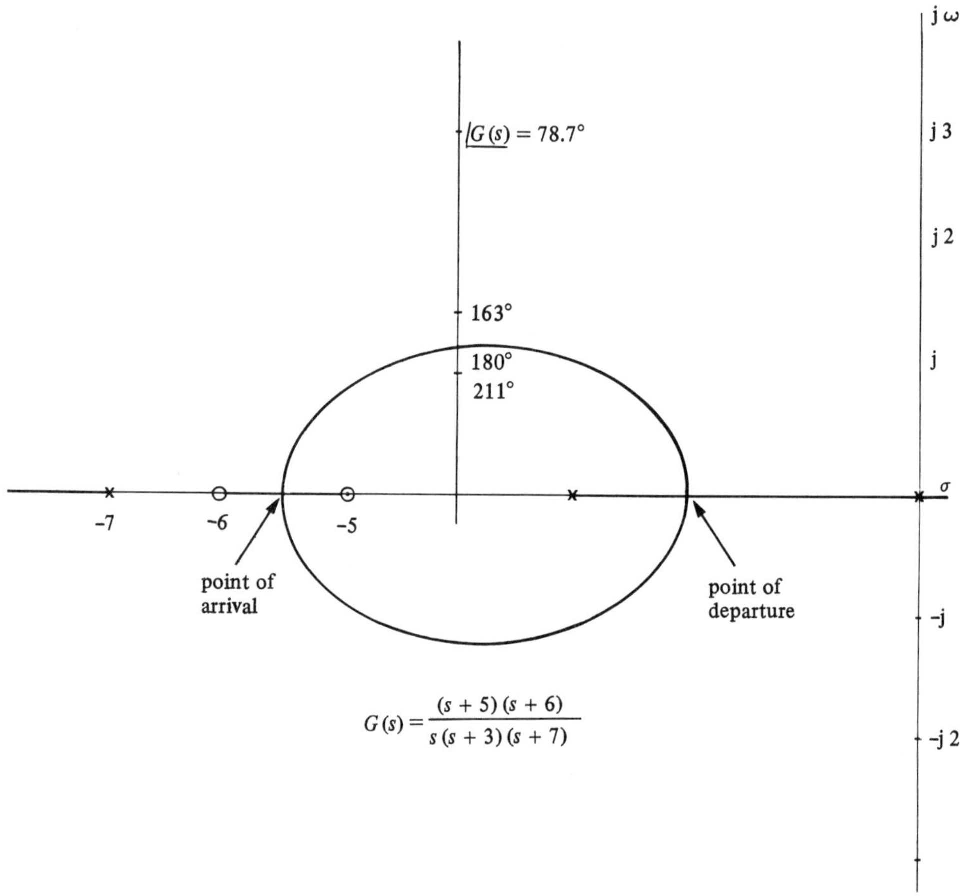

Fig. 5.9 Confirmation of the root locus by use of the angle test.

Each of these cases can be satisfied by application of the magnitude test at a single point. However, a more general solution will require the system transfer function when a particular value of K is selected. This can be achieved if all the branches of the locus are scaled. The points on each branch having the same selected value of K correspond to the system poles (or the system zeros for a zero locus) when that value of K is selected. This can be illustrated in the next example.

Example 5.3. A unity feedback system has a forward gain $A(s)$ given by:

$$A(s) = \frac{5 \times 10^{13} A}{(s + 10^4)(s + 5 \times 10^4)(s + 10^5)} \qquad \ldots (5.25)$$

where A is set by a variable gain control.

Determine the maximum value of A for marginal stability and the values of A which will result in critical damping and a Q factor of 1. In each of these cases, estimate the system closed-loop transfer function.

Solution. The locus for this problem is shown in Fig. 5.10; note that only one half is shown as the other half is the mirror image. The locus was constructed using parts of real axis, asymptotes, point of departure and imaginary axis crossing point. Parts of the real axis were scaled to find the point of departure. Other K values have been found on the real axis beyond the last pole and at the imaginary axis crossing point. Finally, a 60° line has been drawn (Q factor of 1). The point where the locus crosses this line has been checked by the angle test and the corresponding K value has been found. Note that all these K values are equivalent to $5 \times 10^{13} A$.

The maximum value of A occurs when the locus crosses the imaginary axis, i.e. when $K = 97 \times 10^{13}$:

$$\therefore \quad A_{max} = \frac{97 \times 10^{13}}{5 \times 10^{13}} = 19.4 \ .$$

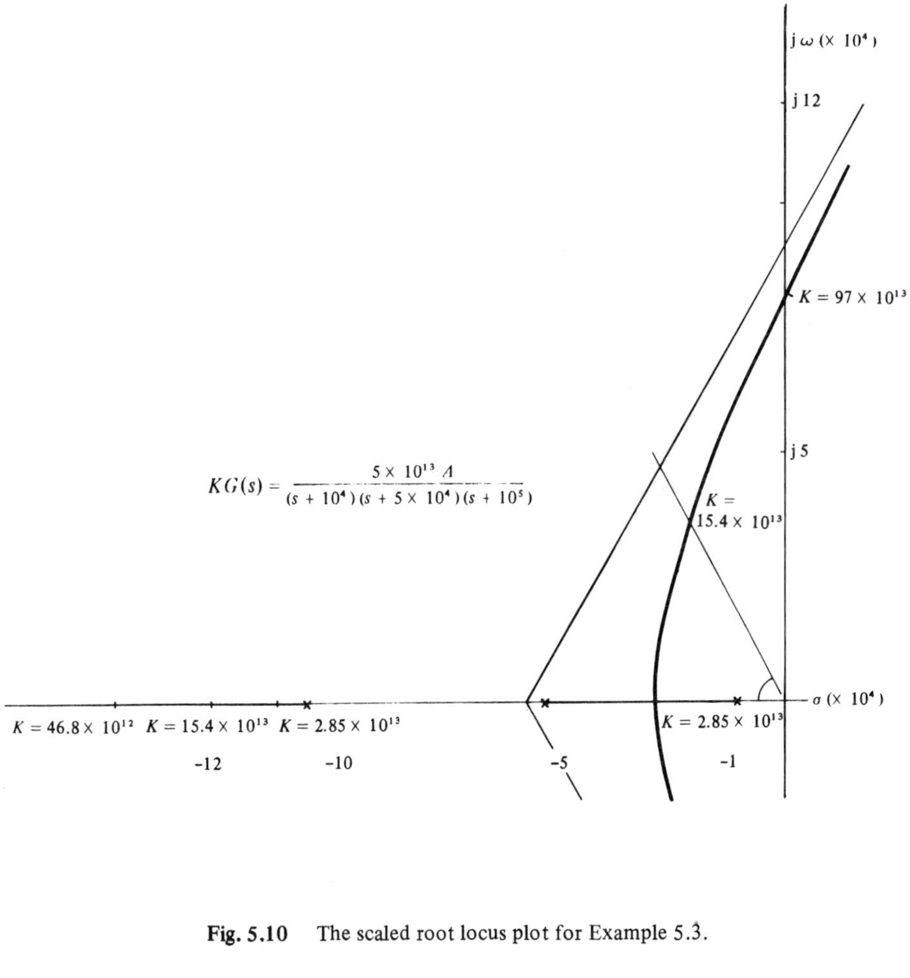

Fig. 5.10 The scaled root locus plot for Example 5.3.

Similarly, for critical damping (point of departure) $K = 2.85 \times 10^{13}$ and $A = \dfrac{2.85}{5} = 0.57$.

For a Q factor of unity $\theta = 60°$, $K = 15.4 \times 10^{13}$ and $A = \dfrac{15.4}{5} = 3.08$.

As the closed loop transfer function is to be estimated, consider the general expression for the closed loop gain $A_f(s)$:

$$A_f(s) = \frac{\dfrac{5 \times 10^{13} A}{(s + 10^4)(s + 5 \times 10^4)(s + 10^5)}}{1 + \dfrac{5 \times 10^{13} A}{(s + 10^4)(s + 5 \times 10^4)(s + 10^5)}} \qquad \ldots (5.26)$$

$$= \frac{5 \times 10^{13} A}{(s + 10^4)(s + 5 \times 10^4)(s + 10^5) + 5 \times 10^{13} A} \; . \qquad \ldots (5.27)$$

The characteristic equation for this expression is a cubic, the roots of which are given by the root locus. A particular set of three roots apply to points on the three branches of the locus having the same values of K (or A).

For the critical damping, two points coincide at $s = -2.7 \times 10^4$ when K is 2.85×10^{13}. On the third branch, K is 2.85×10^{13} at an estimated $s = -10.55 \times 10^4$. The closed loop transfer function is therefore given by:

$$A_f(s) = \frac{2.85 \times 10^{13}}{(s + 2.7 \times 10^4)^2 (s + 10.55 \times 10^4)} \; . \qquad \ldots (5.28)$$

Similarly, for the unity Q factor case, two of the poles lie at $s = -2 \times 10^4 \pm j3.6 \times 10^4$ when $K = 15.4 \times 10^{13}$. The corresponding real axis point where K has the same value is $s = -12 \times 10^4$.

$$\therefore \quad A_f(s) = \frac{15.4 \times 10^{13}}{(s + 12 \times 10^4)(s + 2 \times 10^4 - 3.6 \times 10^4)(s + 2 \times 10^4 + j3.6 \times 10^4)} \; , \qquad \ldots (5.29)$$

or

$$A_f(s) = \frac{15.4 \times 10^{13}}{(s + 12 \times 10^4)(s^2 + 4 \times 10^4 s + 1.7 \times 10^9)} \; . \qquad \ldots (5.30)$$

5.9 COMPLETE PLOTTING OF A SCALED ROOT LOCUS

In this section, two examples will be used to illustrate the entire technique of plotting a scaled locus.

Example 5.4. The transfer function of a system is given by:

$$T(s) = \frac{25 \times 10^7 A\,(s+600)}{s\,(s+1000)\,(s^2+200s+5\times 10^5) + 10^7 A\,(s+600)} \quad \ldots(5.31)$$

Construct a scaled root locus showing how the system pole positions vary with the value of the gain A. Hence determine the maximum value of A for stable operation and the transfer function when A is half this value.

Solution. The scaled locus is shown in Fig. 5.11. The following steps were taken in the construction.
The *CE* was arranged in the root locus form,

$$1 + KG(s) = 1 + \frac{10^7 A\,(s+600)}{s\,(s+1000)\,(s+100-\mathrm{j}700)\,(s+100+\mathrm{j}700)} = 0\,. \quad \ldots(5.32)$$

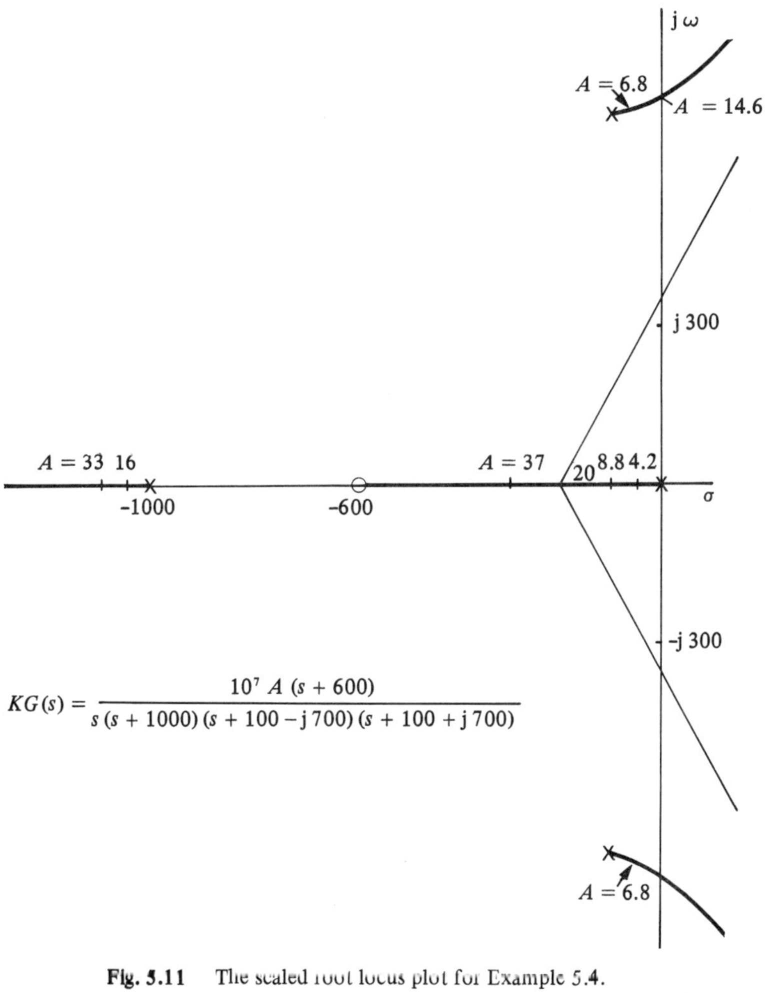

Fig. 5.11 The scaled root locus plot for Example 5.4.

The poles and zeros of $G(s)$ were plotted and the parts of the real axis to the left of an odd number of poles and zeros are marked as part of the locus. The number of asymptotes was found from 4 – 1 = 3. These lie at ±60° and 180° and they meet on the real axis at:

$$s = \frac{0 - 100 - 100 - 1000 + 600}{3} = -200 \,. \qquad \ldots (5.33)$$

The asymptotes were then drawn.

The locus probably goes from the complex poles to the asymptotes at ±60°. To confirm this, the angle of departure from the complex poles was required. Taking a point s_1 near the pole at $s = -100 + j700$,

$$\tan^{-1}(7/5) - (180° - \tan^{-1} 7) - 90° - \tan^{-1}(7/9) - \theta_1 = \pm 180° \qquad \ldots (5.34)$$

or

$$\tan^{-1}(7/5) + \tan^{-1} 7 - 90° - \tan^{-1}(7/9) = \theta_1$$

Hence

$$\theta_1 = 8° \,.$$

The angle test was then applied to points on the imaginary axis. Using the last result suggested a point at $s = j710$. The results were as follows:

$s = j710$	$\angle G(s) = -167°$
$s = j750$	$\angle G(s) = -188°$
$s = j735$	$\angle G(s) = -180.8°$

As this was within one degree, the crossing points of $\pm j735$ were entered.

To scale the locus, the magnitude test was applied at a number of points with the following results:

$s = j735$	$10^7 A = 1.46 \times 10^8$	$A = 14.6$
$s = -50 + j715$	$10^7 A = 6.8 \times 10^7$	$A = 6.8$
$s = -50$	$10^7 A = 4.2 \times 10^7$	$A = 4.2$
$s = -100$	$10^7 A = 8.8 \times 10^7$	$A = 8.8$
$s = -200$	$10^7 A = 2 \times 10^8$	$A = 20$
$s = -300$	$10^7 A = 3.7 \times 10^8$	$A = 37$
$s = -1050$	$10^7 A = 1.6 \times 10^8$	$A = 16$
$s = -1100$	$10^7 A = 3.3 \times 10^8$	$A = 33$

These results give sufficient information within the stable range to estimate the system transfer function for any value of A.

The maximum stable value of A is that obtaining to the imaginary axis crossing point, i.e. 14.6. The estimated points on the locus at which A is 7.3 half A_{max} are $-40 \pm j720$, -75 and -1030. The required transfer function is therefore:

$$T(s) = \frac{25 \times 10^7 \times 7.3\,(s + 600)}{(s + 75)(s + 1030)(s^2 + 80s + 52 \times 10^4)} \,. \qquad \ldots (5.35)$$

Example 5.5. A unity feedback system has an open-loop gain given by the following transfer function.

$$T(s) = K\,\frac{(s+7)\,(s^2+4s+13)}{s\,(s+8)} . \qquad \ldots(5.36)$$

Plot a scaled root locus for the closed-loop system and hence determine the values of K which would result in critical damping.

Solution. The locus is constructed using standard procedures and is shown in Fig. 5.12a. Having plotted the open-loop poles and zeros, the real axis parts of the locus are shown between the origin and $s = -7$ and between $s = -8$ and $-\infty$. As there are three zeros and two poles, there is one asymptote *from* infinity. In addition, there are three branches of the locus, each of which finish at one of the finite zeros.

One possibility is that the locus breaks away from the real axis somewhere between -8 and $-\infty$ and then proceeds directly to the complex zeros. A first check then is to find the angle of arrival at the complex zeros.

$$90° + 31° - 26.6° - (180° - 56.3°) + \theta_1 = 180° . \qquad \ldots(5.37)$$

From which,

$$\theta_1 = -150.7°.$$

This shows that the above suggestion is not correct and that the branches leaving the real axis will break in to the real axis between the origin and -7 and break away again to go to the complex zeros. The magnitude test is applied to points along the real axis and the results are plotted in Fig. 5.12b. This shows a point of inflection at $s = -4.5$. Rules (5.22) and (5.23) give the angles of arrival and departure as shown. Fig. 5.12b also shows a maximum and therefore a point of departure at $s = -10.5$.

To confirm the shape, the angle test is applied along vertical lines at $s = -7$ and $s = -9$ to find two additional points. These and other points are then scaled as shown.

The answer to the problem is that critical damping will occur when $K = 0.092$ (with an additional real pole near the origin) and at $K = 0.41$ with three coincident poles at $s = -4.5$.

5.10 THE ROOT LOCUS FOR POSITIVE FEEDBACK SYSTEMS

The root locus for systems considered so far have been obtained from result (5.8).

$$1 + KG(s) = 0 , \qquad \ldots(5.8)$$

where K has been a positive number between zero and infinity. I sone systems, particularly positive feedback systems, the characteristic equation has the form:

$$1 - KG(s) = 0 , \qquad \ldots(5.38)$$

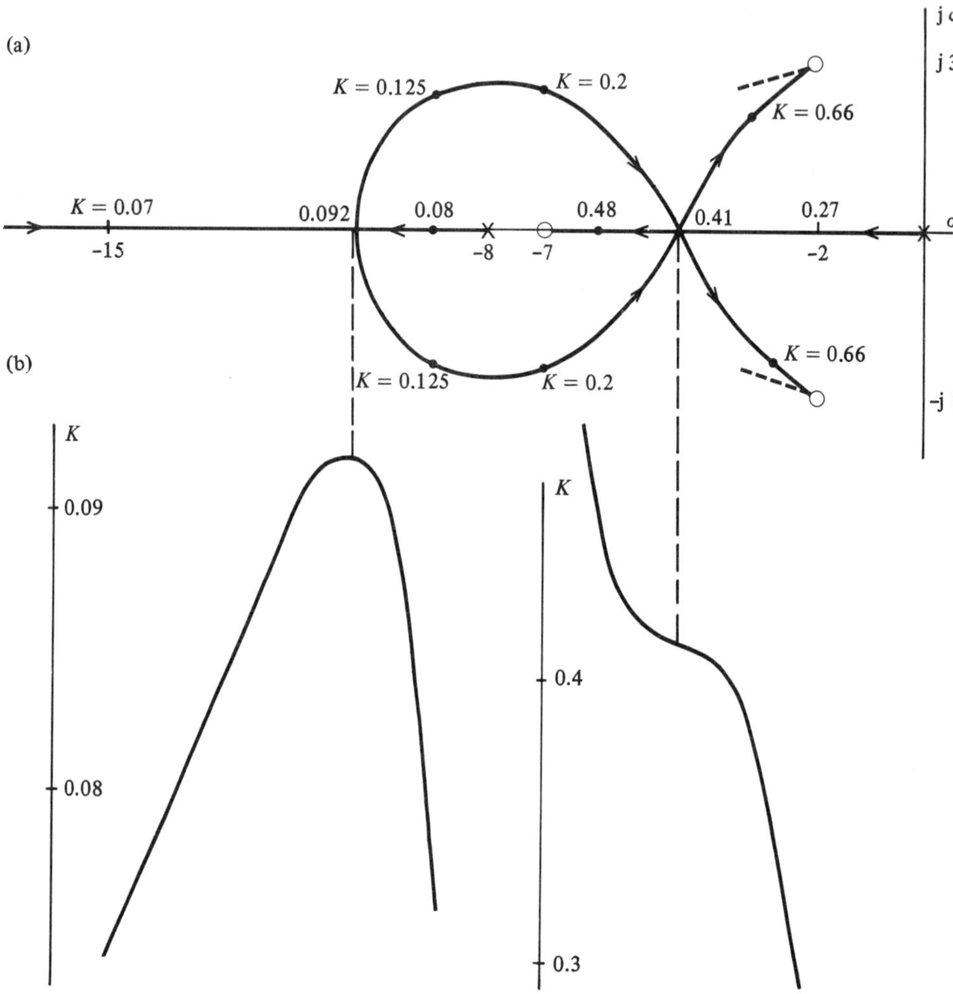

Fig. 5.12 Construction of the scaled root locus for Example 5.5. a) The locus. b) K values for points along the real axis.

where again K is a positive number. Alternatively, in the original form $1 + KG(s) = 0$, K itself can be negative if, for example, an inverting amplifier is involved. In either case, rearrangement yields:

$$KG(s) = 1\angle 0° \qquad \ldots(5.39)$$

This results in a modified angle test. The magnitude test for points on the locus is, however, unchanged. For any point on the locus,

$$\angle G(s) = 0° \pm m360° \qquad \ldots(5.40)$$

and

$$K = \left|\frac{1}{G(s)}\right| \qquad \ldots(5.41)$$

Some of the other rules are changed as a result of these modifications.

i. The parts of the real axis to the left of an *even* number of poles and zeros is part of the locus. (Even includes zero.) ...(5.42)

ii. The angles of the asymptotes to or from infinity are $\dfrac{n360°}{n_p - n_z}$ . ...(5.43)

This gives the following angles:

$n_p - n_z$	*Angles*
1	0°
2	0°, 180°
3	0°, ±120°
4	0°, ±90°, 180°
5	0°, ±72°, ±144°

iii. The asymptote meeting point rule is unchanged. ...(5.44)

iv. The angles of arrival and departure for complex poles and zeros are found in the same way as for negative feedback systems, but the total angle will now be zero. ...(5.45)

All the other techniques discussed earlier in the chapter are still applicable.

Example 5.6. A Wien Bridge *CR* network has a voltage transfer function given by:

$$T(s) = \frac{100s}{s^2 + 300s + 10^4} \ .$$

It is used with a non-inverting summing amplifier as shown in Fig. 5.13. Plot a root locus for the system to determine the value of A which would enable the system to be used as a selective amplifier.

Solution. Referring to Fig. 5.13,

$$V_o = A\ T(s) V' \ . \qquad ...(5.46)$$

$$V' = V_{in} + V_o \ . \qquad ...(5.47)$$

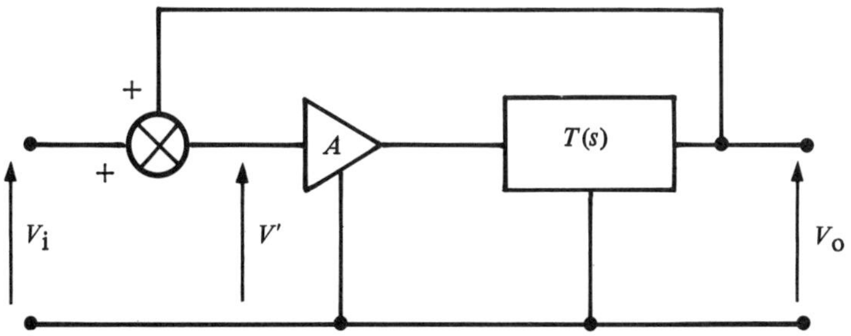

Fig. 5.13 The positive feedback circuit for Example 5.6.

Hence, the system transfer function is given by:

$$A_f(s) = \frac{A\,T(s)}{1 - A\,T(s)} \quad \ldots(5.48)$$

This is in the positive feedback root locus form but it will be convenient to substitute for $T(s)$ and then to rearrange the result.

$$A_f(s) = \frac{\dfrac{100As}{s^2 + 300s + 10^4}}{1 - \dfrac{100As}{s^2 + 300s + 10^4}} \quad \ldots(5.49)$$

$$= \frac{100As}{s^2 + 100\,(3 - A)\,s + 10^4} \quad \ldots(5.50)$$

The poles or denominator roots could obviously be found by solving the quadratic for different values of A, but instead we will use the root locus technique working from the denominator of expression (5.49). The constructed locus is shown in Fig. 5.14 and the following steps were used.

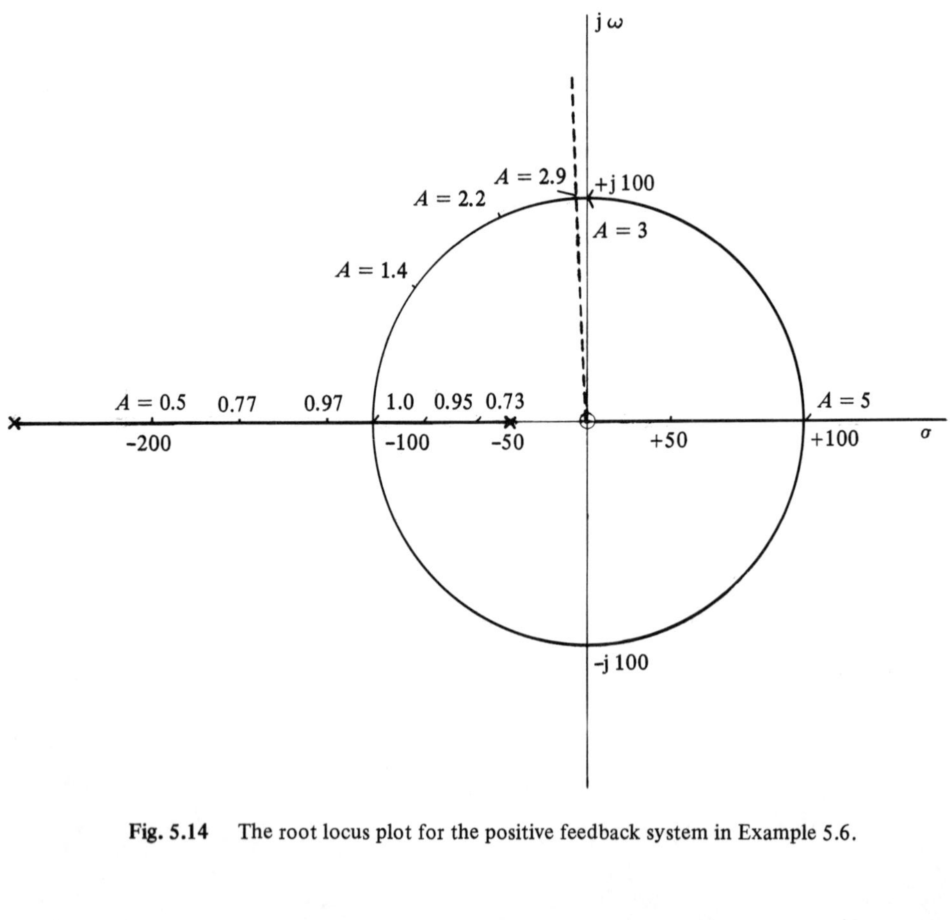

Fig. 5.14 The root locus plot for the positive feedback system in Example 5.6.

$G_{(s)}$ is, in this case, the original transfer function of the *CR* network which has a zero at the origin and poles at –38.2 and –261.8. These are plotted on the *s* plane as shown. From rule (5.42), all positive parts of the real axis are parts of the locus and so is the part between –38.2 and –261.8. There is only one asymptote to infinity, at 0°. Scaling along the real axis between –38.2 and –261.8 shows a maximum *K* of 100 at $s = -100$. As *K* is $100A$, the corresponding *A* value is 1. Other *A* values are shown along this real axis part of the locus. Scaling along the positive part of the real axis shows a minimum *A* of 5 at $s = +100$.

Applying the angle test along the imaginary axis shows a crossing point at $s = \text{j}100$ with a corresponding *A* of 3. The remainder of the locus is obviously the circle shown.

A selective amplifier would require a *Q* factor of at least 10. The corresponding angle θ with respect to the negative real axis, is 87°. An additional point along this line is checked at $s = -5 + \text{j}100$. The corresponding *A* is 2.9 and a system with this value would therefore provide the required selective amplifier. It is interesting to note, however, that only a small change in gain would be sufficient to make the system unstable.

Example 5.7. A positive feedback system has an open-loop gain given by:

$$T(s) = \frac{20A}{(s+5)(s^2+4s+13)}$$

If one tenth of the output is added to the input, draw a root locus and hence determine the maximum value of *A* for stable operation. State the system transfer function if *A* is adjusted to half this maximum value.

Solution. From the data given, the characteristic equation for the system is:

$$1 - \frac{2A}{(s+5)(s+2-\text{j}3)(s+2+\text{j}3)} = 0 \; ; \qquad \ldots(5.51)$$

or

$$2AG_{(s)} = 1\angle 0° \; .$$

The poles of $G(s)$ are at –5, –2 + j3 and –2 – j3; these are plotted in Fig. 5.15. The real axis to the right of the real pole is part of the locus shown. There are three asymptotes at ±120° and 0° and these meet on the real axis at $s = \frac{-5-2-2}{3} = -3$.

Writing the equation for the angle of departure from the complex poles,

$$0° - 90° - 45° - \theta = 0° \; , \qquad \ldots(5.52)$$

and

$$\theta = -135° \; .$$

Angle tests along a line $s = -3.25$ shows that a point $-3.25 + \text{j}2.5$ is on the locus. The scaling for this point and a few others is shown on the diagram. The values are all for $2A$; the maximum *A* is therefore 65/2 or 32.5. Notice that the instability resulting from greater values of *A* would be a simple exponent to infinity.

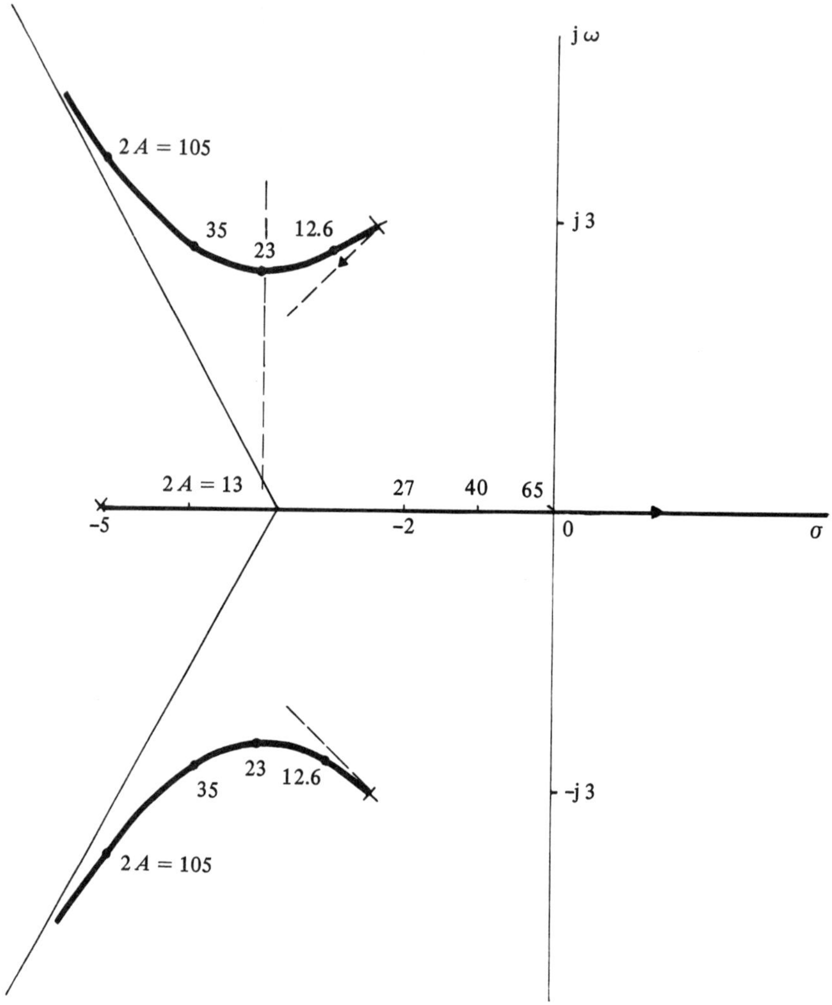

Fig. 5.15 Construction of the root locus for the positive feedback system in Example 5.7.

5.11 SUMMARY OF STEPS FOR PLOTTING THE ROOT LOCUS

i. Arrange *CE* in root locus form $1 \pm KG(s) = 0$.
ii. Plot poles and zeros of $G(s)$.
iii. Show real axis parts of the locus. Rules (5.19) or (5.42).
iv. Plot asymptotes to or from infinity. Rules (5.17), (5.18) or (5.43), (5.44).
v. Determine points of arrival and departure on the real axis. Rules (5.20), (5.21) and possibly (5.22), (5.23).
vi. Determine angles of arrival or departure from complex poles or zeros. Rules (5.24) or (5.45).

vii. Determine the crossing point of the j axis. Rules (5.21) or (5.39).
viii. Confirm further detail using the angle test. Rules (5.10) or (5.39).
ix. Scale as required using rules (5.13) or (5.41).

General. Although the reasons for the plotting of root loci have been introduced, this chapter has been almost entirely concerned with the techniques of locus plotting and scaling. While mastering these techniques, it is easy to lose sight of the objective, which is the prediction of system behaviour under transient or steady-state conditions. The locus allows us to obtain the system poles and zeros for a particular value of the parameter under investigation or, to choose the parameter value to provide particular pole or zero positions. In either case, the resulting system transfer function can be used as described in Chapter 3 to determine the transient response or as in Chapter 4 to investigate the steady-state response. These, and other applications are studied in more detail in the next chapter.

5.12 EXAMPLES FOR FURTHER PRACTICE

Example 5.8. A second-order control system with unity feedback has a transfer function given by: $T(s) = \dfrac{K/J}{s^2 + sF_v/J + K/J}$ Construct the loci, by calculation, of the system poles obtained by variation of a) K, b) F_v and c) J with, in each case, the other two terms constant. Take the constant values, as required, as $K = 5$, $F_v = 0.1$ and $J = 0.04$.

Example 5.9. The characteristic equations of three systems are given below. In each case, rearrange the equations in the 'root locus form' and determine the poles and zeros of $G(s)$.

a) $4s^2\,(1 + K) + 16s + 80 = 0$.

b) $s^4 + 8s^3 + (7 + K)\,s^2 + 2Ks + 10K = 0$.

c) $s^3 + s^2\,(4 + 3K) + 4s + 75K = 0$.

(2 zeros at 0, poles at −2 + j4 and −2 − j4: zeros at −1 + j3 and −1 −j3, poles at 0, 0, −1 and −7: zeros at +j5 and −j5, poles at 0, −2 and −2.)

Example 5.10. A number of characteristic equations have been rearranged in the form $1 + KG(s) = 0$. The resulting expressions for $G(s)$ are listed below. In each case, sketch the general form of the root locus and construct a fully scaled root locus. As a check, the answers given with each example are, respectively, imaginary axis crossing points and points of arrival and departure from the real axis. In each case, the values of s and K are given.

a) $\dfrac{(s + 4)}{s\,(s + 2)\,(s + 6)}$, (–; −1.07, 1.67).

b) $\dfrac{(s+12)}{(s+1)(s+3)(s+4)}$, (7.35, 35; –1.83, 0.207).

c) $\dfrac{1}{s(s+100)(s+120)}$, (110, 2.6 × 10^6; –36, 1.95 × 10^5).

d) $\dfrac{1}{(s+0.1)(s+0.8)(s+1.2)(s+2)}$, (0.77, 2.6; –0.36, 0.158, –1.72, 0.217).

e) $\dfrac{(s+20)(s+45)}{s(s+10)(s+50)}$, (–; –5.9, 1.935; –30, 80).

f) $\dfrac{1}{(s+100)(s^2+400s+53\times 10^4)}$, (755, 2.3 × 10^8).

g) $\dfrac{1}{(s+1)(s^2+16s+68)}$, (9.2, 1371; –7.7, 27.4, –3.63, 60.74).

h) $\dfrac{s^3}{(s+50)(s+120)(s+300)}$, (62, 13.8; –177, 1.6).

i) $\dfrac{1}{(s+2000)(s^2+8000s+4.1\times 10^7)}$, (7550, 4.9 × 10^{11}).

j) $\dfrac{s^2+1600s+8.9\times 10^5}{s(s+450)(s+1200)}$, (–; –248, 85.97).

k) $\dfrac{s}{(s+45\,000)(s^2+49\times 10^6)}$, (–; –9050, 5.2 × 10^8; –19 500, 5.6 × 10^8).

l) $\dfrac{(s+4)}{s(s+1)(s+20)(s+50)}$, (28, 5.5 × 10^4; –0.53, 69.1).

m) $\dfrac{(s+2)}{s(s+1)(s+20)(s+50)}$, (30.5, 6.5 × 10^4; –0.58, 165; –4, 4416; –7.7, 4709).

n) $\dfrac{(s+22)}{(s+1)(s+24)(s+26)}$, (–; –24.9, 8.15).

o) $\dfrac{(s+23)}{(s+1)(s+24)(s+26)}$, (–; 24.8, 12.7; –20.9, 150; –14.4, 173.5).

p) $\dfrac{1}{s(s+10)(s^2+10s+41)}$, (4.5, 2445; –5, 400; –2.9, 420; –7.1, 420).

q) $\dfrac{1}{s(s+10)(s^2+10s+89)}$, (6.7, 6470; –5, 1600; –5 ± j4.4, 1980).

r) $\dfrac{s(s+1-j2)(s+1+j2)}{(s+3-j10)(s+3+j10)}$, (2.38, 9.1; 9.8, 0.06).

Example 5.11. A number of characteristic equations have been rearranged in the form $1 - KG(s) = 0$. The resulting expressions for $G(s)$ are listed below. In each case, sketch the general form of the root locus, and then construct a fully scaled root locus. As a check the answers given with each example are respectively, imaginary axis crossing points and points of arrival and departure from the real axis. In each case, the values of s and K are given.

a) $\dfrac{(s+2)}{(s+1)(s+3)(s+5)}$, (0, 7.5).

b) $\dfrac{s}{(s+20)(s+40)}$, (28.2, 60; –29, 3.41; 28, 116.6).

c) $\dfrac{1}{(s+100)(s+500)(s+600)}$, ($0, 3 \times 10^7$; –552, 1128).

d) $\dfrac{(s-100)}{(s^2+250s+55\,625)}$, (285, 251; +400, 1052).

6

Root Locus Applications

6.1 INTRODUCTION

Chapter 5 was concerned with developing the techniques of root locus plotting. While concentrating on these techniques, it is very easy to lose sight of the original objectives and it may be useful to reconsider these before examining specific applications.

The transient and steady-state responses of systems can be predicted if the system transfer function is known.

The system transfer function can be defined in terms of the poles and zeros of the function.

The transient and steady-state responses can be predicted directly in terms of these poles and zeros.

The s plane positions of the poles and zeros of the transfer function depend upon the values of the system components and parameters.

Changes in the values of components or parameters will 'move' the poles and/or the zeros of the function.

A root locus plot shows how the system poles (or zeros) move when a particular component or parameter is varied between zero and infinity.

A root locus can therefore be used to predict variation in steady-state and transient response as a particular parameter is varied.

In this chapter, we shall examine a number of problems for which root locus plots provide an essential part of the solution. The details of the plotting procedure will not be given unless there are any special difficulties or points of interest. The reader, however, would be well advised to plot the locus for himself with additional detail if required.

The first two examples are concerned with appreciation of what the root locus represents. For the modification of the transient response, a control system is examined; in the second example, a feedback amplifier is used to demonstrate changes in the steady-state response. The next examples make use of locus plots in the determination of the poles and zeros of a higher-order transfer function and in the choice of components so that particular pole positions can be obtained. Other examples include compensation for both control systems and feedback amplifiers, sinusoidal oscillators and active filters.

6.2 A UNITY FEEDBACK CONTROL SYSTEM

Example 6.1. A unity feedback position control system has an open-loop transfer function given by:

$$T_0(s) = \frac{K(s+1)}{s(s+0.5)(s+4)(s+10)} \quad \ldots (6.1)$$

The constant K can be selected as required. By means of a root locus, investigate the step response for values of K between five times that required to give critical damping and up to 0.7 times the value which will result in instability.

Solution. The scaled root locus is shown in Fig. 6.1 with points scaled for values of K. The locus was constructed using standard techniques including application of the angle test to a number of points in the complex region to find the exact path of the locus. Reading from the diagram, the limiting values of K are 3.1 for critical damping and 459 for unstable operation. We shall therefore investigate the step response for K values of 15, 130 and 320. The corresponding closed-loop transfer functions are obtained from the original open-loop function and from the root locus.

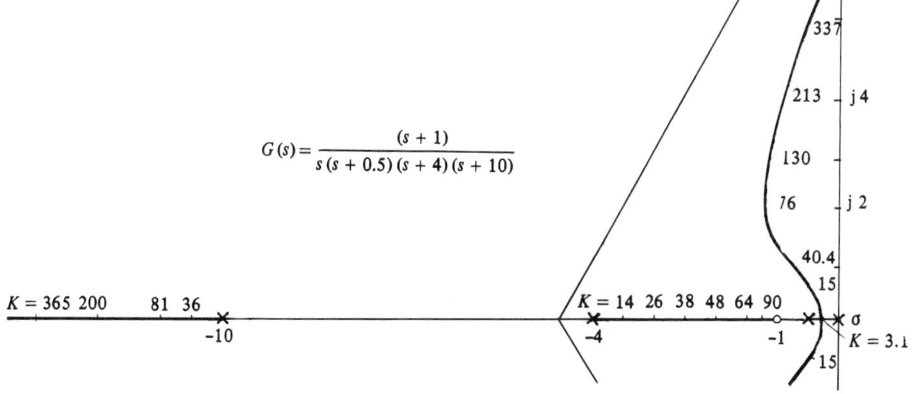

Fig. 6.1 The root locus for the unity feedback position control system in Example 6.1.

$$T_c(s) = \frac{\dfrac{K(s+1)}{s(s+0.5)(s+4)(s+10)}}{1 + \dfrac{K(s+1)}{s(s+0.5)(s+4)(s+10)}} \qquad \ldots(6.2)$$

$$= \frac{K(s+1)}{s(s+0.5)(s+4)(s+10) + K(s+1)} \; . \qquad \ldots(6.3)$$

This system has a zero at $s = -1$ and four poles whose values depend upon the value of K. If a particular value of K can be found as a scaling factor on each of the four branches of the locus, these points are the required pole positions for that value of K. For example, when K is 15, the points on the complex branches are $(-0.4 \pm j05)$; on the two real branches we can estimate the other two pole positions with $K = 15$ as -3.4 and -10.2.

The first of the required transfer functions is therefore given by:

$$T_{c1}(s) = \frac{15(s+1)}{(s+3.4)(s+10.2)(s+0.4-j0.5)(s+0.4+j0.5)} \; . \qquad \ldots(6.4)$$

The transfer functions for the other two values of K can be found in the same way:

$$T_{c2}(s) = \frac{130(s+1)}{(s+1.15)(s+11.4)(s+0.95-j3)(s+0.95+j3)} \; , \qquad \ldots(6.5)$$

and

$$T_{c3}(s) = \frac{320(s+1)}{(s+1.05)(s+12.8)(s+0.35-j4.9)(s+0.35+j4.9)} \; . \qquad \ldots(6.6)$$

For the step response, we must now multiply each transfer function by $1/s$ and we can then use the pole zero diagrams and the methods described in Chapter 3.

Examining the pole zero diagrams in Fig. 6.2, we can notice that in cases (b) and (c), the pole and zero near to $s = -1$ virtually cancel each other. We shall also see that the high frequency pole will make only a very small contribution to the response and can probably be ignored, particularly in case (a). Consider first case (b); the response $C(s)$ is given by:

$$C(s) \triangleq \frac{130}{s(s+11.4)(s+0.95-j3)(s+0.95+j3)} \qquad \ldots(6.7)$$

$$= \frac{A}{s} + \frac{B}{s+11.4} + \frac{C}{s+0.95-j3} + \frac{C}{s+0.95+j3} \; .$$

Working from the pole zero diagram, we find that:

$$A = 1.15, \quad B = -0.1, \quad \text{and} \quad C = 0.63\underline{/-124^\circ -90^\circ} \; .$$

Hence,

$$c(t) = 1.15 - 0.1e^{-11.4t} + 1.26e^{-0.95t}\sin(3t - 124^\circ) \; . \qquad \ldots(6.8)$$

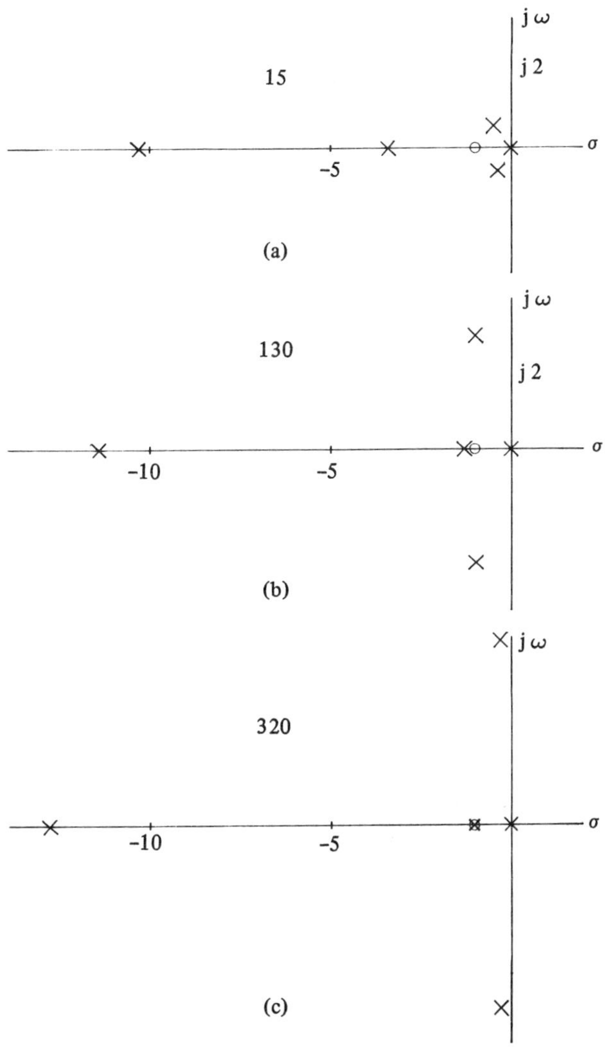

Fig. 6.2 Step response pole zero diagrams for the unity feedback position control system in Example 6.1. a) $K = 15$. b) $K = 130$. c) $K = 320$.

Using the same methods, case (c) results in:

$$c(t) = 1.04 - 0.14e^{-12.8t} + 0.994e^{-0.35t} \sin(4.9t - 115°) , \qquad \ldots(6.9)$$

and case (a) results in:

$$c(t) = 1.05 + 0.168e^{-3.4t} - 0.02e^{-10.2t} + 1.228e^{-0.4t} \sin(0.5t - 101°) . \qquad \ldots(6.10)$$

Note that in each of these results, the steady-state response is not exactly unity which it should be for this type of system. This is the result of the approximation made in cancelling the pole and the zero and in errors due to estimation of the pole positions from the locus. The three response curves are shown in Fig. 6.3 for comparison with the pole zero diagrams and the original root locus.

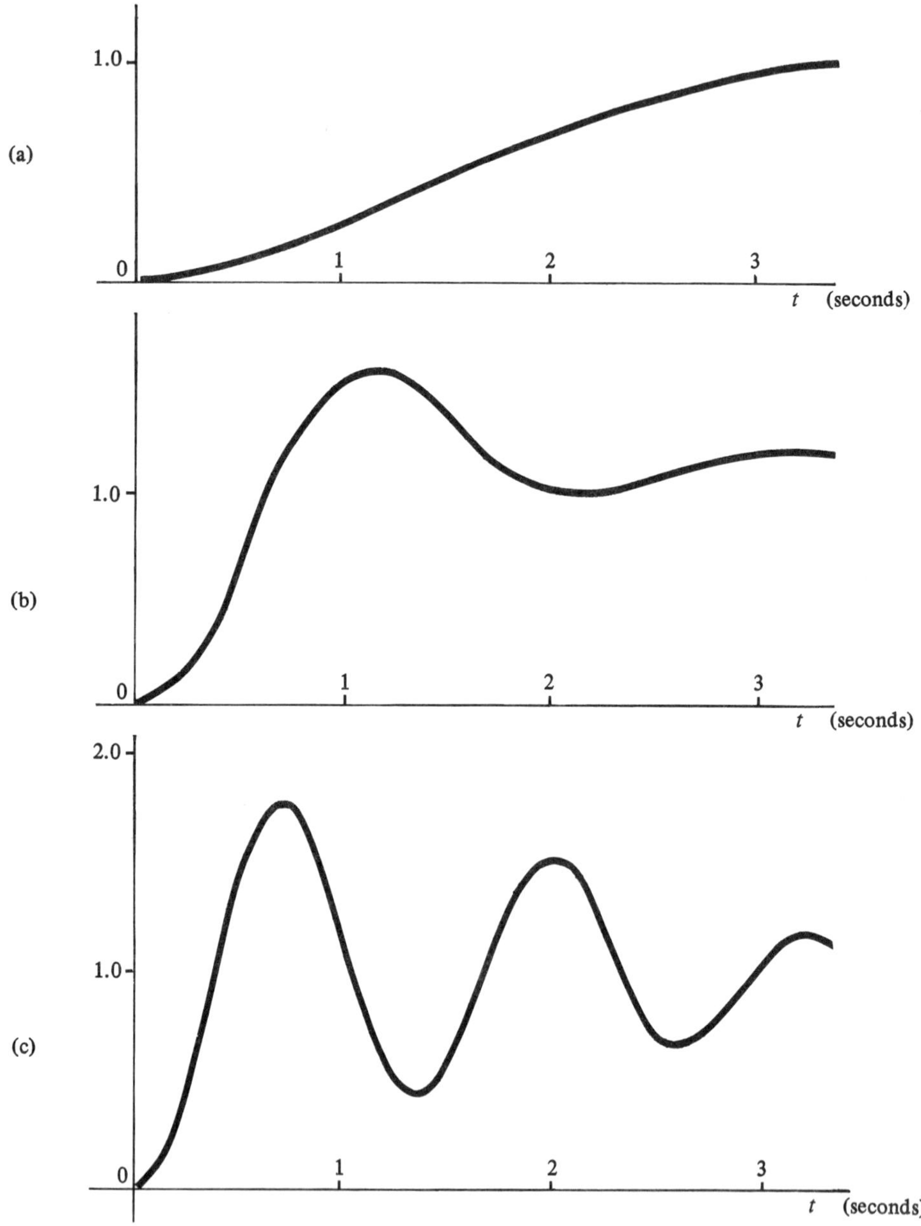

Fig. 6.3 The step response for the unity feedback control system in Example 6.1. a) $K = 15$. b) $K = 130$. c) $K = 320$.

6.3 THE FREQUENCY RESPONSE OF A FEEDBACK AMPLIFIER

Example 6.2. The gain of a three-stage d.c. amplifier is given by:

$$A(j\omega) = \frac{-800}{(1 + j\omega/\omega_1)(1 + j\omega/\omega_2)(1 + j\omega/\omega_3)} . \qquad \ldots(6.11)$$

The break frequencies ω_1, ω_2 and ω_3 are respectively 10^6, 2.5×10^6 and 3×10^6 rads^{-1}. Feedback, negative at low frequencies, is applied with a network having a transfer function β, where β is a real fraction. With the aid of a root locus, find the maximum value of β for stable operation. Find also, the system transfer function when a) there are two coincident real poles and b) when there is a complex conjugate pair of poles giving a Q factor of 5. In each case, draw Bode plots for comparison with the root locus.

Solution. The gain in the s domain, $A(s)$, is obtained by substituting s for $j\omega$.

Hence,

$$A(s) = \frac{-800}{(1 + s/\omega_1)(1 + s/\omega_2)(1 + s/\omega_3)} . \qquad \ldots (6.12)$$

Multiplying numerator and denominator by $\omega_1 \omega_2 \omega_3$,

$$A(s) = \frac{-800\omega_1 \omega_2 \omega_3}{(s + \omega_1)(s + \omega_2)(s + \omega_3)} . \qquad \ldots (6.13)$$

Applying the standard feedback formula $\dfrac{A}{1 - \beta A}$;

$$A_f(s) = \frac{\dfrac{-800\omega_1 \omega_2 \omega_3}{(s + \omega_1)(s + \omega_2)(s + \omega_3)}}{1 + \dfrac{800\beta\omega_1 \omega_2 \omega_3}{(s + \omega_1)(s + \omega_2)(s + \omega_3)}} . \qquad \ldots (6.14)$$

We could now multiply through by $(s + \omega_1)(s + \omega_2)(s + \omega_3)$ to obtain a cubic for the denominator whose roots depend upon β. The denominator of (6.14) is in the correct form for a root locus plot, which will provide us with the required roots for any chosen value of β. Substituting the values for the break frequencies into the denominator of (6.14);

$$1 + \frac{800 \times 7.5 \times 10^{18}\beta}{(s + 10^6)(s + 2.5 \times 10^6)(s + 3 \times 10^6)} = 0 . \qquad \ldots (6.15)$$

The scaled root locus is shown in Fig. 6.4. The scaling factors shown are for β, which is given by: $\beta = \dfrac{K}{6 \times 10^{21}}$.

The maximum value of β will be that which takes the locus through the imaginary axis. From the scaled locus, this value is 0.0129 and would result in oscillation at $\beta = 3.62 \times 10^6$ rads^{-1}.

a) When $\beta = 1.26 \times 10^{-4}$, there are two coincident real poles at $s = -1.57 \times 10^6$. On the remaining real branch, the same value of β, 1.26×10^{-4}, occurs at $s = -3.37 \times 10^6$. Referring to expressions (6.14) and (6.15), the system transfer function is thus given by:

$$A_f(s) = \frac{-6 \times 10^{21}}{(s + 1.57 \times 10^6)^2 (s + 3.37 \times 10^6)} . \qquad \ldots (6.16)$$

From which,

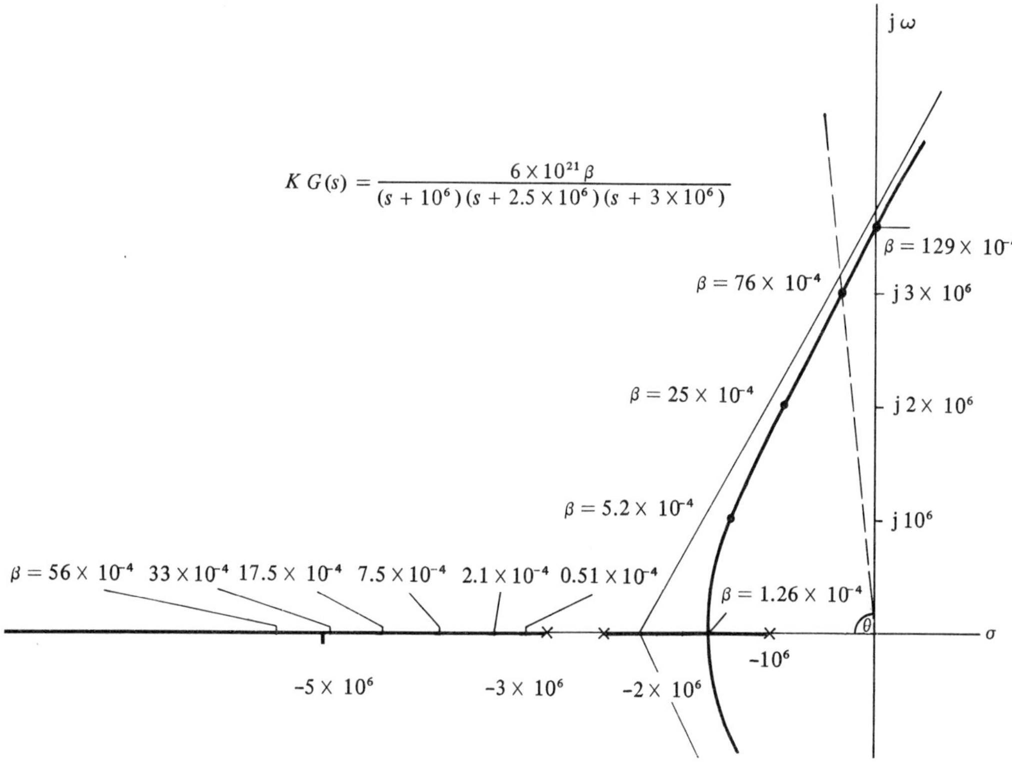

Fig. 6.4 The root locus for the three stage feedback amplifier in Example 6.2.

$$A_f(j\omega) = \frac{-722}{(1+j\omega/\omega_a)^2\,(1+j\omega/\omega_b)}, \qquad \ldots(6.17)$$

where ω_a and ω_b are 1.57×10^6 and 3.37×10^6 rads^{-1} respectively.

b) For a Q factor of 5, $\dfrac{1}{2\cos\theta} = 5$ or $\theta = 84°$. The corresponding point on the locus is $s = (-0.3 + j3)\,10^6$ when β is 7.6×10^{-3}. The point on the real axis branch with the same β is at $s = -5.8\times10^6$. Again referring to (6.14) and (6.15), the system transfer function is given by:

$$A_f(s) = \frac{6\times10^{21}}{(s+5.8\times10^6)(s+0.3\times10^6 - j3\times10^6)(s+0.3\times10^6+j3\times10^6)};$$

$$A_f(s) = \frac{6\times10^{21}}{(s+5.8\times10^6)(s^2+0.6\times10^6 s+9.1\times10^{12})}. \qquad \ldots(6.18)$$

From which,

$$A(j\omega) = \frac{1.72\times10^9}{j\omega\left(1+\dfrac{j\omega}{5.8\times10^6}\right)\left(\dfrac{j\omega}{6\times10^5}+1-\dfrac{j1.5\times10^7}{\omega}\right)}. \qquad \ldots(6.19)$$

The two Bode gain plots are shown in Fig. 6.5 with the necessary constructional details.

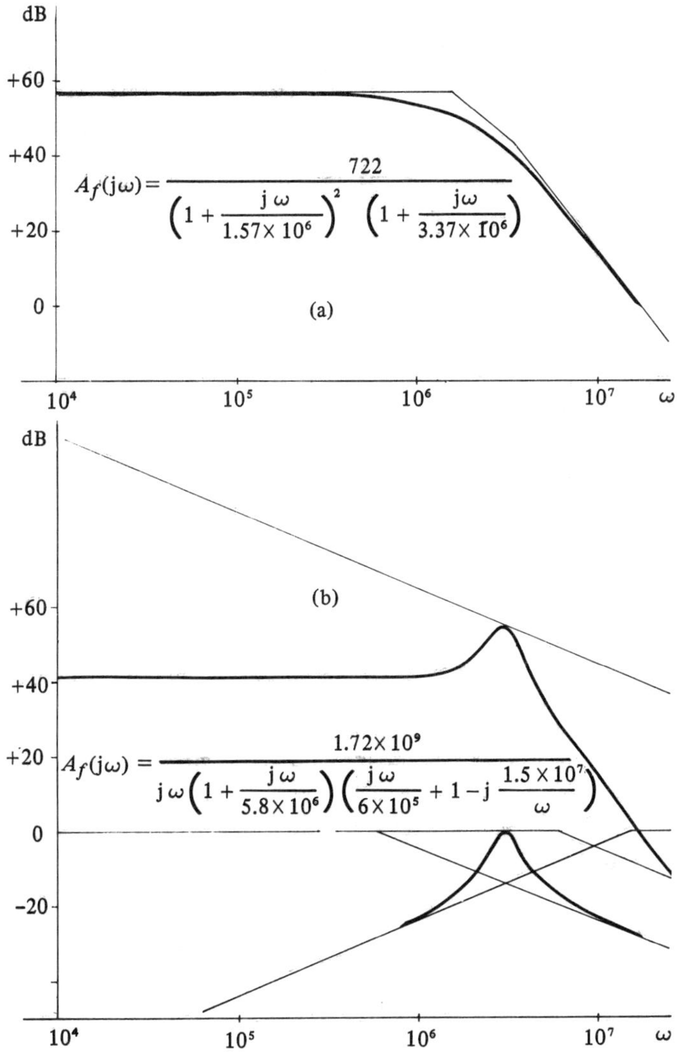

Fig. 6.5 The Bode gain plot for the three stage feedback amplifier in Example 6.2. a) When there are to coincident real poles. b) When there are two complex conjugate poles having a Q factor of 5.

6.4 FACTORIZING A POLYNOMIAL BY USE OF THE ROOT LOCUS

The next example is concerned with the transfer function of a passive network, the denominator of which is a third-order polynomial. A root locus is used to determine the poles of this transfer function, i.e. to factorize the polynomial.

Example 6.3. The phase shift network shown in Fig. 6.6 has equal resistors and capacitors of 10 kΩ and 1 μF respectively. Determine the poles and zeros of the voltage transfer function.

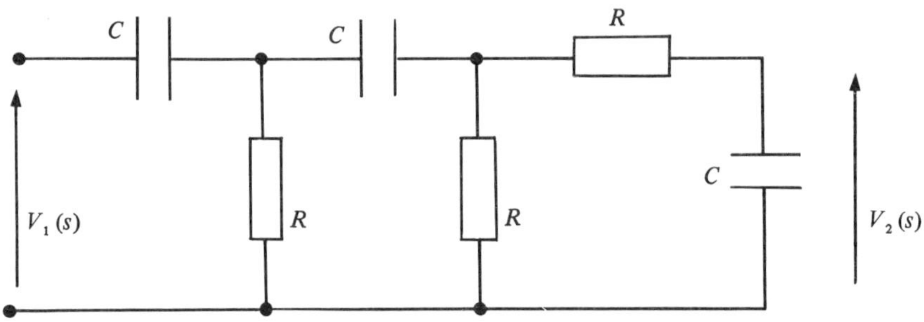

Fig. 6.6 The *RC* phase shift network for Example 6.3.

Solution. The network may be solved using mesh or nodal analysis to find the required transfer function:

$$T(s) = \frac{s^2/CR}{s^3 + \dfrac{6s^2}{CR} + \dfrac{5s}{C^2R^2} + \dfrac{1}{C^3R^3}} \quad . \qquad \ldots(6.20)$$

Inserting values,

$$T(s) = \frac{100s^2}{s^3 + 600s^2 + 50\,000s + 10^6} \quad . \qquad \ldots(6.21)$$

The poles of this function can only be found if the cubic denominator can be factorized. This can be done by the *partition method* using a root locus plot. The characteristic equation is given by:

$$s^3 + 600s^2 + 50\,000s + 10^6 = 0 \quad . \qquad \ldots(6.22)$$

This can be rewritten in one of two forms for partition. The first of these expressed in root locus form is:

$$1 + \frac{10^6}{s^3 + 600s^2 + 50\,000s} = 0 \quad . \qquad \ldots(6.23)$$

The denominator can now be factorized;

$$1 + \frac{K}{s\,(s+100)\,(s+500)} = 0 \quad . \qquad \ldots(6.24)$$

This expression has been written in the general root locus form with the constant 10^6 replaced by K. If a scaled root locus is now plotted for (6.24) the points on the three branches where K is 10^6 give the required roots of the cubic in (6.22). A sketch of the essential parts of the locus is shown in Fig. 6.7a. Only the real parts of the locus need to be scaled as the roots for this passive network must be real. The three points producing a K of 10^6 are –31, –64 and –505.

The required transfer function is therefore given by:

$$T(s) = \frac{100s^2}{(s+31)\,(s+64)\,(s+505)} \quad . \qquad \ldots(6.25)$$

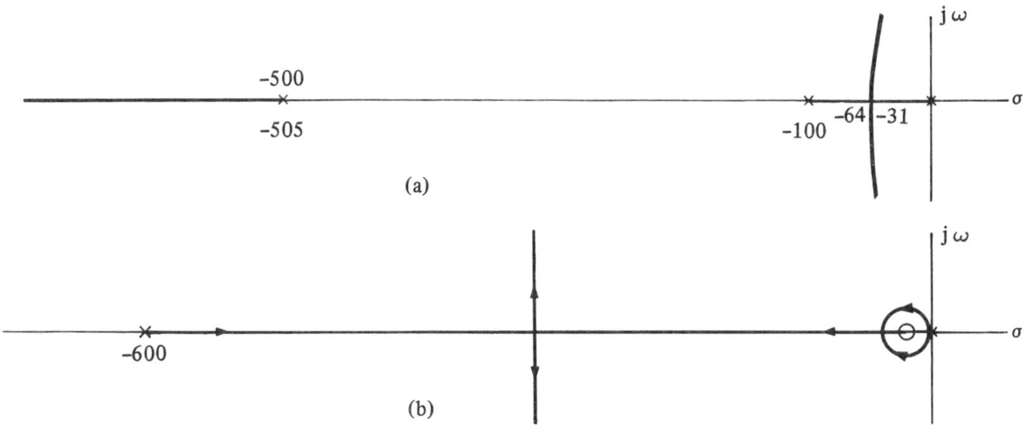

Fig. 6.7 Alternative root locus plots for determination of the transfer function poles in Example 6.3.

The poles are at –31, –64 and –505 and there is a double zero at the origin.

The second (alternative) partition can be obtained from expression (6.22) as follows:

$$1 + \frac{50\,000\,(s + 20)}{s^2\,(s + 600)} \qquad \text{. . . (6.26)}$$

The form of the locus is shown in Fig. 6.7b and it is left to the reader to confirm that this gives the same results.

Higher order polynomials can be treated in a similar manner. Fourth- and fifth-order polynomials can be factorized with the aid of a single partition and one root locus. The partition of a sixth-order polynomial will itself produce a cubic which must be factorized before the root locus can be constructed. This factorization can, of course, be achieved by the use of a separate root locus.

6.5 CHOICE OF COMPONENT VALUE BY ROOT LOCUS PLOT

If an electrical network is to be used to provide a specific function such as a resonant frequency or a Q factor and one component can be conveniently adjusted to this end, a root locus can be constructed, taking this component as the parameter of the locus. The scaling of the locus can then be used to select the optimum value of the component.

Example 6.4. The bridged T network shown in Fig. 6.8 can be used to provide complex poles and complex zeros. Investigate the range of C required and find the position of the zeros when the poles result in a unity Q factor.

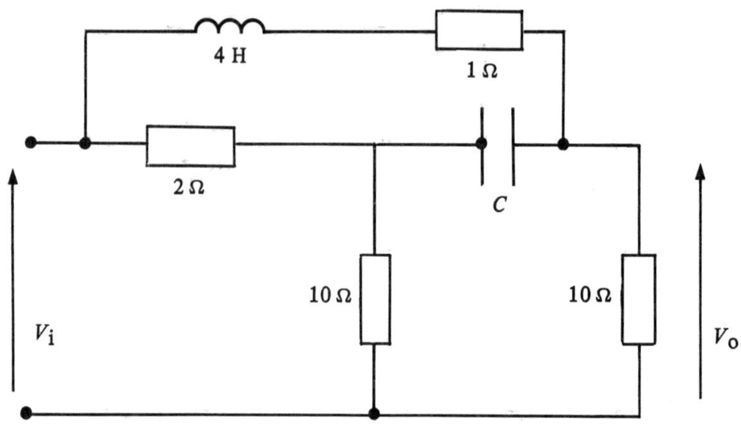

Fig. 6.8 The bridged *T* network having complex poles and complex zeros for Example 6.4.

Solution. If nodal analysis is used, the simplified equations can be expressed:

$$0.5 V_i = (0.6 + sC)\, V_a - sCV_o \,, \qquad \ldots(6.27)$$

$$V_i = -sCV_a + (1.1 + 4s + sC + 4s^2C)\, V_o \,. \qquad \ldots(6.28)$$

Solving by determinants and collecting terms yields:

$$T(s) = \frac{V_o}{V_i} = \frac{2s^2C + 1.5sC + 0.6}{2.8s^2C + 0.24s + 1.7sC + 0.66} \,. \qquad \ldots(6.29)$$

Both numerator and denominator of (6.29) are quadratic and have some coefficients which are functions of C. Thus two separate root locus plots will be required, one for the zeros and one for the poles. For the zero locus, from (6.29):

$$2s^2C + 1.5sC + 0.6 = 0 \qquad \ldots(6.30)$$

rearranging,

$$0.6 + 2Cs\,(s + 0.75) = 0 \qquad \ldots(6.31)$$

or,

$$1 + 3.33Cs\,(s + 0.75) = 0 \qquad \ldots(6.32)$$

Expression (6.32) is in the root locus form for a parameter $3.33C$ where $G(s)$ has two zeros and no finite poles. The required locus is shown in Fig. 6.9a with scaling for values of C.

For the pole locus, from (6.29):

$$2.8s^2C + 0.24s + 1.7sC + 0.66 = 0 \,, \qquad \ldots(6.33)$$

rearranging,

$$0.24s + 0.66 + 2.8Cs\,(s + 0.607) = 0 \,, \qquad \ldots(6.34)$$

or,

$$\frac{1 + 11.67\, Cs\,(s + 0.607)}{(s + 2.75)} = 0 \,. \qquad \ldots(6.35)$$

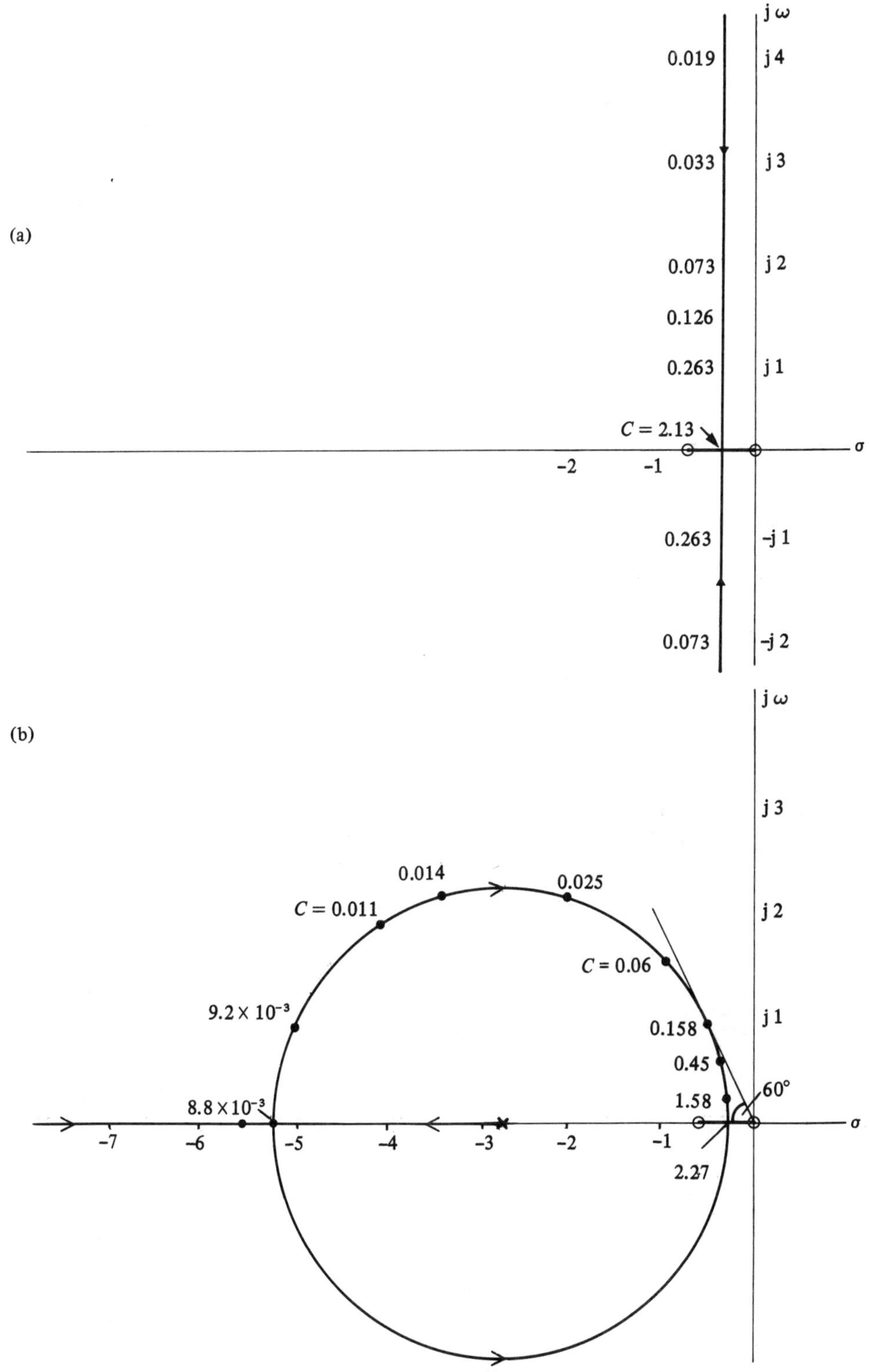

Fig. 6.9 Root locus plots for the bridged *T* network in Example 6.4. a) The zero locus. b) The pole locus.

Expression (6.35) is in the required root locus form where $11.67C$ is the parameter and $G(s)$ has two zeros and one pole. The resulting pole locus is shown in Fig. 6.9b.

The following conclusions can be drawn from Fig. 6.9.

i. The zeros of $T(s)$ will be complex for all values of C less than 2.13 F.
ii. The poles of $T(s)$ will be complex for C less than 2.27 F and greater than 8.8×10^{-3} F.
iii. With unity Q factor poles ($0 = 60°$), C is 0.158 F and the transfer function is given by:

$$T(s) = \frac{0.71\,(s + 0.375 - \text{j}1.35)\,(s + 0.375 + \text{j}1.35)}{(s + 0.6 - \text{j}1.06)\,(s + 0.6 + \text{j}1.06)}; \qquad \ldots(6.36)$$

or,

$$T(s) = \frac{0.71\,(s^2 + 0.75s + 3.64)}{(s^2 + 1.2s + 1.48)}. \qquad \ldots(6.37)$$

6.6 THE SENSITIVITY OF AN ACTIVE FILTER TO COMPONENT VARIATION

A root locus plot can be used to investigate the effect of component or parameter variation upon the poles or zeros of a transfer function. This is illustrated in the next example which is concerned with an active low pass filter.

Example 6.5. The circuit shown in Fig. 6.10 is an active low pass filter. It is designed to have a second-order Chebyshev response with a 0.5 dB pass band ripple, a pass band gain of 20 dB and a cut-off frequency of 10 kHz. The amplifier has a non-inverting gain of 10 and the input and output impedances can be assumed to be ideal. Determine the system transfer function, the frequency response and, by means of root locus plots, investigate the sensitivity of the system poles to variation in gain or component values.

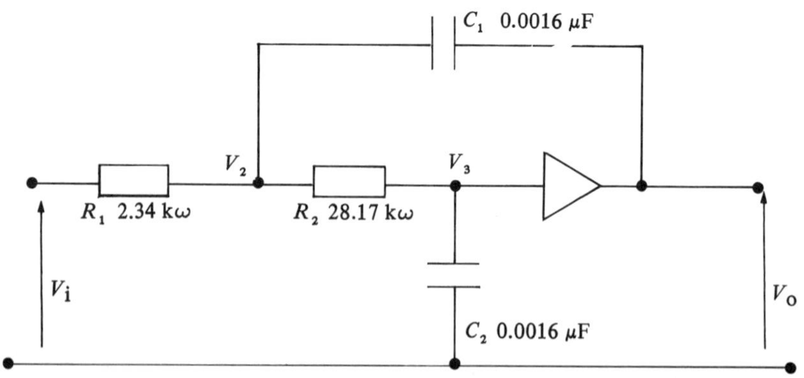

Fig. 6.10 The active low pass filter circuit for Example 6.5.

Solution. The transfer function can be obtained by writing nodal equations:

$$(V_i - V_2)\,G_1 = V_2\,(G_2 + sC_1) - V_3\,(G_2 + sC_1A)\ , \qquad \ldots(6.38)$$

also,

$$V_2 = V_3\,(1 + sC_2R_2)\ . \qquad \ldots(6.39)$$

Rearranging and solving for V_3/V_i;

$$\frac{V_3}{V_i} = \frac{G_1}{s^2C_1C_2R_2 + s\left[C_2(1 + R_2/R_1) + C_1(1 - A)\right] + G_1}\ . \qquad \ldots(6.40)$$

Dividing through by $C_1C_2R_2$, writing $G_1 = 1/R_1$, and putting $V_o = AV_3$:

$$T(s) = \frac{\dfrac{A}{C_1C_2R_1R_2}}{s^2 + s\left[\dfrac{1 + R_2/R_1}{C_1R_2} + \dfrac{1 - A}{C_2R_2}\right] + \dfrac{1}{C_1C_2R_1R_2}}\ . \qquad \ldots(6.41)$$

Substituting the given values:

$$T(s) = \frac{5.93 \times 10^{10}}{s^2 + 8.96 \times 10^4 s + 5.93 \times 10^9}\ , \qquad \ldots(6.42)$$

for which the poles lie at $s = -4.48 \times 10^4 \pm j6.26 \times 10^4$.

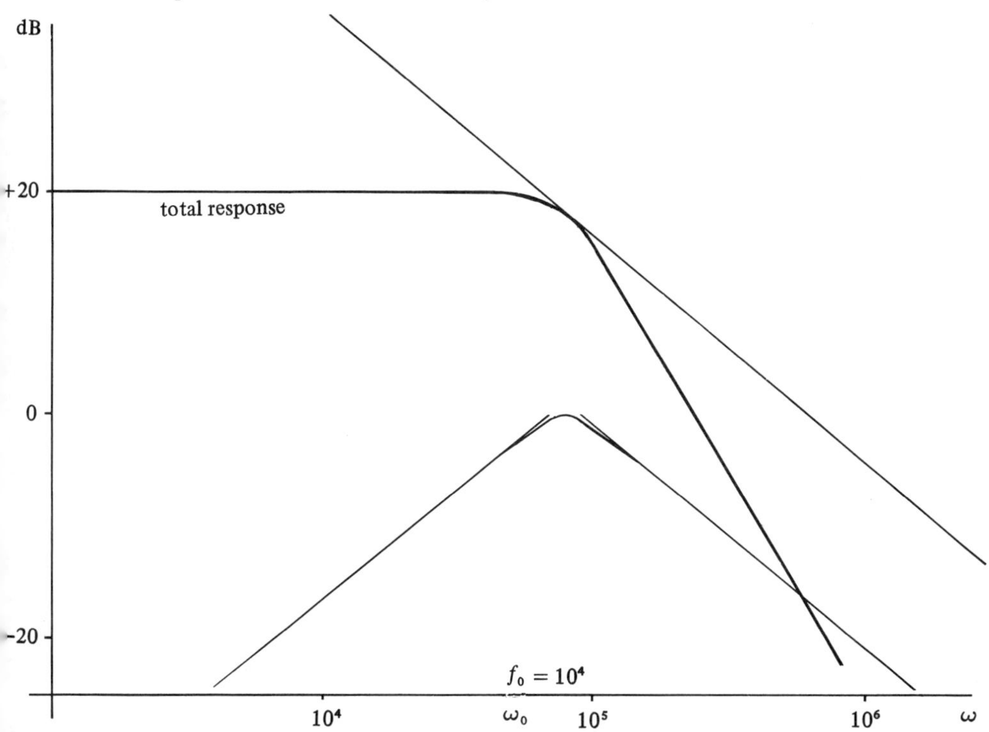

Fig. 6.11 Construction of the Bode gain plot for the active low pass filter circuit in Example 6.5.

To obtain the frequency response, substitute $s = j\omega$ and rearrange,

$$T(j\omega) = \frac{6.62 \times 10^5}{j\omega\left(\dfrac{j\omega}{8.96 \times 10^4} + 1 + \dfrac{6.62 \times 10^4}{j\omega}\right)} \; . \qquad \ldots(6.43)$$

The Bode gain frequency response is constructed using the techniques described in Chapter 4 with the result shown in Fig. 6.11.

Root locus plots are used to investigate the effects of gain and component variation. The characteristic equation must therefore be rearranged as required. If, for the first investigation, the gain A is to be the parameter, the characteristic equation becomes:

$$s^2 + s\left[\frac{1 + R_2/R_1}{C_1R_2} + \frac{1}{C_1C_2}\right] + \frac{1}{C_1C_2R_1R_2} - \frac{As}{C_2R_2} = 0 \; . \qquad \ldots(6.44)$$

Inserting values for C_1, C_2, R_1 and R_2 and arranging in the root locus form:

$$1 - \frac{2.22 \times 10^4 As}{(s + 2.04 \times 10^4)(s + 2.9 \times 10^5)} = 0 \; . \qquad \ldots(6.45)$$

This root locus has the positive feedback form with $2.22 \times 10^4 A$ as the parameter. The locus is shown in Fig. 6.12a with the scaling shown for values of A. Variation of gain from 7.2 to 14.1 moves the poles from critical damping to unstable operation. As a measure of sensitivity, we can note that a 41 per cent increase in gain will move the poles through an angle of 36° or approximately a 1 per cent change in gain for a 1° shift of pole position.

The corresponding root locus forms when other components are taken as the parameter are for R_2, C_1 and C_2 respectively:

for R_2
$$1 - \frac{5 \times 10^9}{R_2} \; \frac{(s - 3.34 \times 10^4)}{s(s + 2.67 \times 10^5)} \; . \qquad \ldots(6.46)$$

for C_1
$$1 + \frac{4.63 \times 10^{-4}}{C_1} \; \frac{(s + 2.05 \times 10^4)}{s(s - 2 \times 10^5)} \; , \qquad \ldots(6.47)$$

for C_2
$$1 - \frac{3.2 \times 10^{-4}}{C_2} \; \frac{(s - 3 \times 10^4)}{s(s + 2.89 \times 10^5)} \; . \qquad \ldots(6.48)$$

The determination of the expression involving R_1 is left for the reader. The root locus plots for expressions (6.46), (6.47) and (6.48) are shown in Fig. 6.12. It is interesting to note that they all have a similar form despite the fact that both negative and positive feedback types are represented. In each case, a change of about 40 per cent would result in unstable operation or again, a 1 per cent change in the parameter moves the poles through approximately 1°.

Although the two capacitors are equal, we cannot plot a locus using the common value C as the parameter as both C and C^2 will appear in equation (6.41). However, the effect of a change in the common value will be simply a change of scale. The Q factor will be unchanged, but the cut-off frequency will increase as C is reduced.

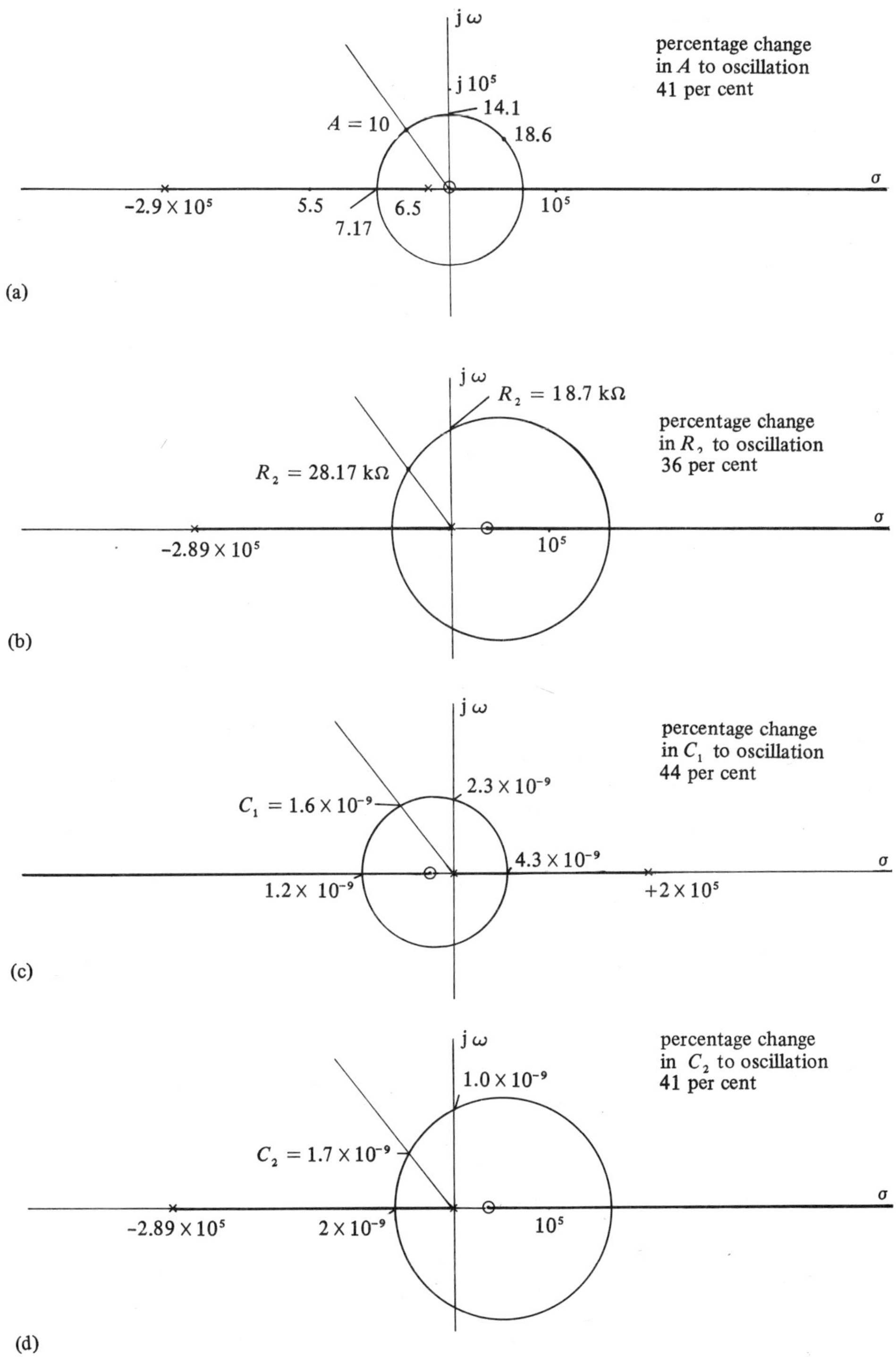

Fig. 6.12 Root locus plots showing the sensitivity of pole positions to variation in circuit parameters in Example 6.5.

6.7 CONTROL SYSTEM COMPENSATION

The next two examples will illustrate the use of root locus plots in the study of control system improvement by means of compensation.

Example 6.6. A type one unity feedback control system has two open-loop time constants of 0.5s and 0.143s. When the gain is adjusted to provide an acceptable maximum overshoot of 16 per cent in the step response, the speed of the response is too slow. As a result, a lead compensator having a transfer function of $\frac{(s+\omega_1)}{(s+5\omega_1)}$ is used in the forward path to improve the response. By means of suitable root locus plots, determine an optimum value for ω_1.

Solution. From the information given, the forward transfer function of the basic system is given by:

$$T(s) = \frac{K_1}{s(s+2)(s+7)}; \qquad \ldots(6.49)$$

and the transfer function for the compensated system is given by:

$$T(s) = \frac{K_a(s+\omega_1)}{s(s+2)(s+7)(s+5\omega_1)}. \qquad \ldots(6.50)$$

Since the system has unity feedback, expressions (6.49) and (6.50) are in the form $KG(s)$ for root locus plotting. As the optimum value for ω_1 is to be investigated, plots are required for the basic system and for several values of ω_1, preferably on the same graph. These plots are shown in Fig. 6.13. Fig. 6.13a shows the pole zero pattern and the real parts of the locus for five different situations. These are respectively, the basic system, $\omega_1 = 1$, $\omega_1 = 1.5$, $\omega_1 = 2.5$ and $\omega_1 = 4$. The corresponding K values are shown as K_1 to K_5. The real axis parts of the locus and the points of departure are also indicated in each case.

Fig. 6.3b shows only the complex parts of the loci for all five cases. These are labelled for convenience by the K numbers at the points of departure and at the imaginary axis crossing points.

The first point to notice from Fig. 6.13a is that when the compensator zero is less than the system pole at $s = -2$, there is a complete branch between this zero and the origin. This means that there will be a real closed-loop pole, near the origin, whatever value of K is used. However, the compensator zero is also a closed-loop system zero and this will tend to minimize the effect of the pole. When the compensator zero is placed beyond the first finite pole and K is large enough to give complex branches, the real pole will be beyond the zero and have a negligible effect upon the step response.

Now turning to Fig. 6.13b; at first it might be supposed that the higher the value of ω_1, the faster would be the step response. But when the percentage overshoot requirement is taken into account, we arrive at a different conclusion.

The percentage overshoot for a second order system is given by $100e^{\frac{-\zeta\pi}{\sqrt{1-\zeta^2}}}$. For a 16 per cent overshoot, this corresponds to a damping factor ζ, of 0.5 or a 60° position for the complex pole. This 60° line is shown and labelled $\zeta = 0.5$. Proceeding from the basic

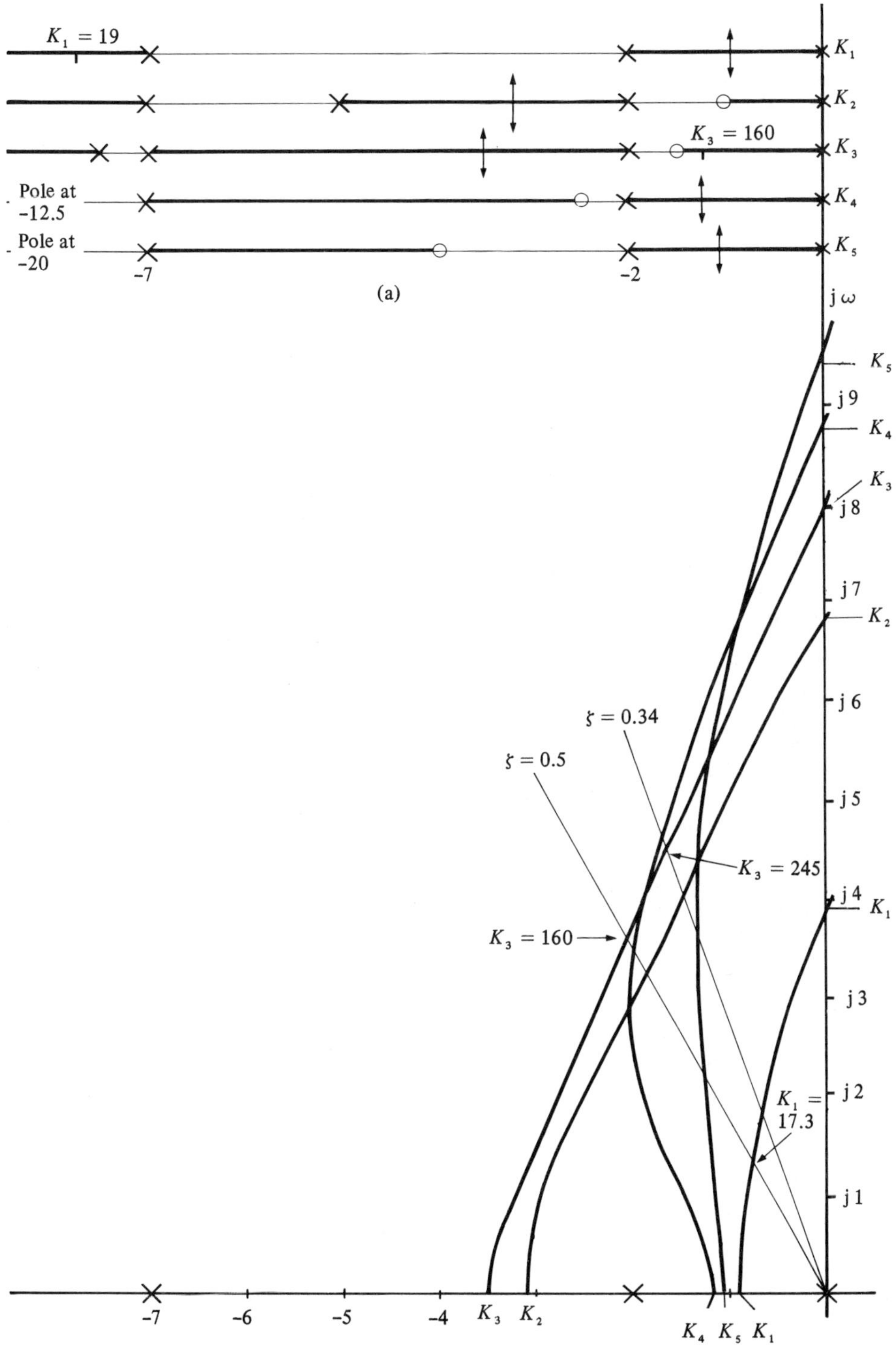

Fig. 6.13 Root locus plots for a compensated unity feedback position control system; the effect of choice of compensator pole and zero.

system K_1 to K_2 and K_3, we can see that the corresponding points occur at higher frequencies (jω) and at greater σ values suggesting a faster response. The result of a further increase in ω_1 to the K_4 and K_5 loci show both the frequency and σ falling again. The K_3 plot appears to be the optimum and the value of K_3, 160, is indicated on the locus. Scaling along the real axis shows that the additional pole occurs at $s = -1.3$.

For comparison purposes, the closed-loop transfer functions for the original system and the compensated system are respectively:

$$T_1(s) = \frac{17.3}{(s + 0.75 - \text{j}1.3)(s + 0.75 + \text{j}1.3)(s + 7.7)} \; . \qquad \ldots(6.51)$$

$$T_3(s) = \frac{160\,(s + 1.5)}{(s + 2 - \text{j}3.5)(s + 2 + \text{j}3.5)(s + 1.3)(s + 11.2)} \; . \qquad \ldots(6.52)$$

The effect of the high frequency pole at $s = -11.2$ is very small and it is convenient to neglect this term in the calculation of the step response. The resulting modified transfer function is given by:

$$T_3(s) = \frac{14.3\,(s + 1.5)}{(s + 2 - \text{j}3.5)(s + 2 + \text{j}3.5)(s + 1.3)} \; . \qquad \ldots(6.53)$$

The step response in each case is obtained in the usual way, by multiplying the transfer function by $1/s$ and, after partial fraction expansion, taking the inverse transform.

The resulting step responses are shown in Fig. 6.14a for $T_1(s)$ and Fig. 6.14b for $T'_3(s)$. It should be noted that the overshoot in the compensated case is only 10 per cent and that the time to this first overshoot is about 0.4 of that in the basic system. The overshoot is less than that expected from the 60° position of the complex poles as a result of the real pole at $s = -1.3$. (The expression for overshoot in terms of damping factor quoted above is only applicable to a second-order system.) This suggests that a further increase in gain could be acceptable. A line corresponding to $\zeta = 0.34$ is shown on Fig. 6.13. The reader could complete this investigation by taking K_2, K_3 and K_4 along this line, to see whether the corresponding overshoot was acceptable and if a further improvement in speed was obtainable. K_4 is unlikely to be satisfactory as the real pole is no longer near the origin and, as a result, the overshoot will be too large. Both K_2 and K_3 will result in a higher frequency oscillatory component but the overshoot will depend upon the real pole and the system zero.

The step response resulting from a K_3 of 245 is shown in Fig. 6.14b and the overshoot is about 22 per cent. This suggests that a K_3 of about 200 will meet the specification and this could be confirmed by the reader. The procedure is to scale along the branches of the K_3 locus to find the points where K_3 is 200. The closed-loop transfer function is then obtained and hence the step response.

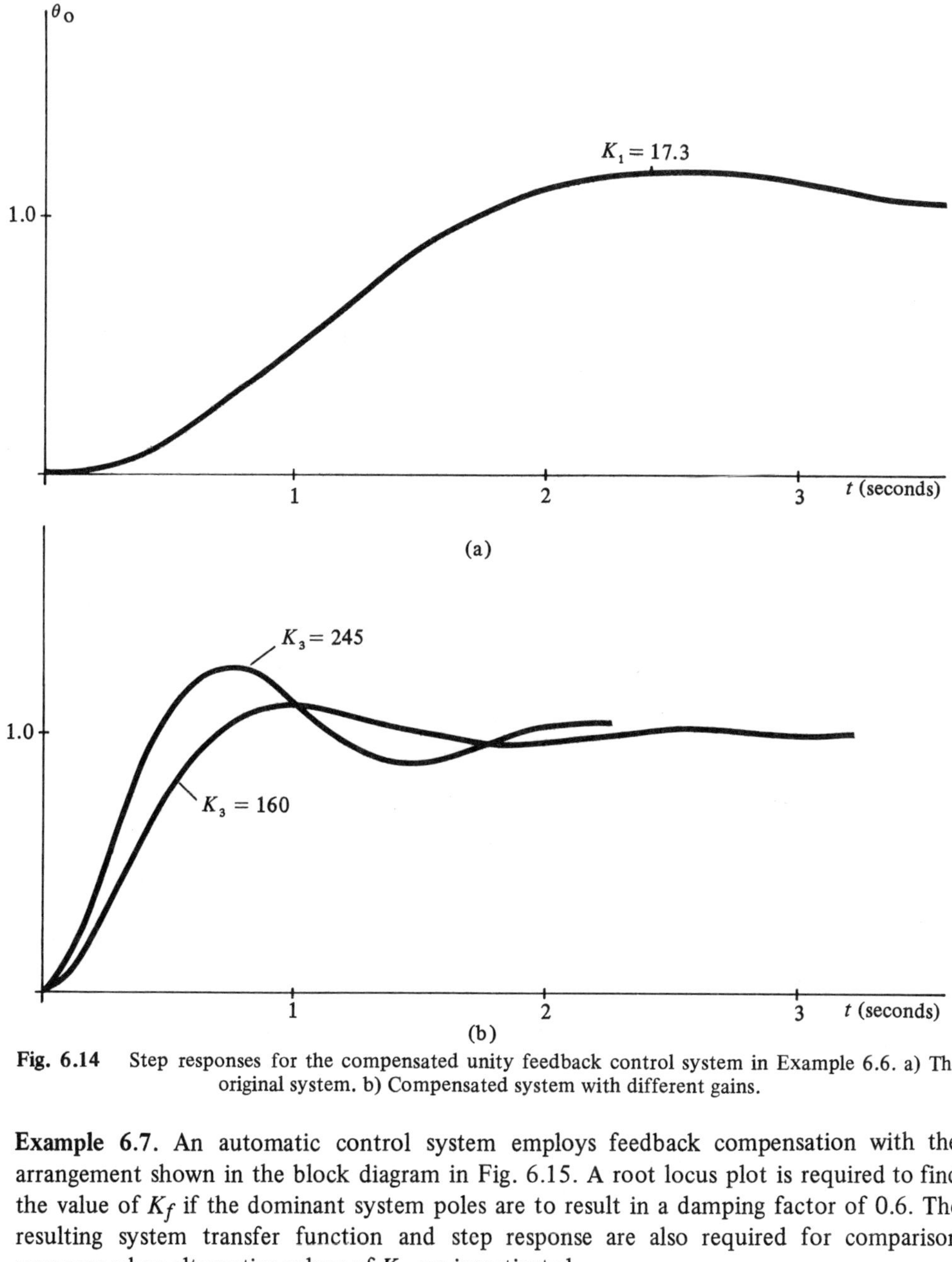

Fig. 6.14 Step responses for the compensated unity feedback control system in Example 6.6. a) The original system. b) Compensated system with different gains.

Example 6.7. An automatic control system employs feedback compensation with the arrangement shown in the block diagram in Fig. 6.15. A root locus plot is required to find the value of K_f if the dominant system poles are to result in a damping factor of 0.6. The resulting system transfer function and step response are also required for comparison purposes when alternative values of K_a are investigated.

Solution. This problem is initially one of manipulation of the block diagram components so that the system transfer function $\frac{C}{R}(s)$ can be arranged in the required root locus form. It is convenient if the block transfer functions are first converted into the pole zero form:

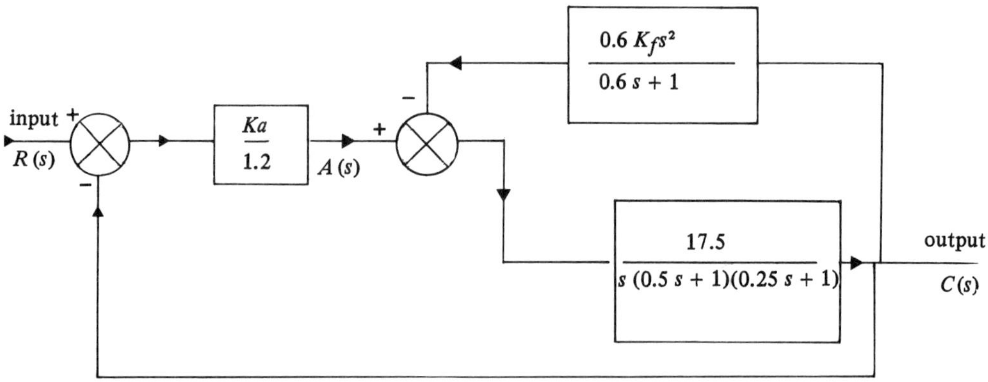

Fig. 6.15 The block diagram for the automatic control system in Example 6.7.

$$\frac{17.5}{s(0.5s+1)(0.25s+1)} = \frac{140}{s(s+2)(s+4)}\ ; \qquad \ldots(6.54)$$

$$\frac{0.6s^2K_f}{0.6s+1} = \frac{K_f s^2}{s+1.67}\ . \qquad \ldots(6.55)$$

Consider next the inner loop:

$$\frac{C}{A}(s) = \frac{\dfrac{140}{s(s+2)(s+4)}}{1 + \dfrac{140K_f s^2}{s(s+2)(s+4)(s+1.67)}}\ , \qquad \ldots(6.56)$$

$$= \frac{140(s+1.67)}{s(s+2)(s+4)(s+1.67) + 140K_f s^2}\ . \qquad \ldots(6.57)$$

Now for the complete system;

$$\frac{C}{R}(s) = \frac{\dfrac{168(s+1.67)}{s(s+2)(s+4)(s+1.67)+140K_f s^2}}{1 + \dfrac{168(s+1.67)}{s(s+2)(s+4)(s+1.67)+140K_f s^2}}\ ; \qquad \ldots(6.58)$$

$$= \frac{168(s+1.67)}{s(s+2)(s+4)(s+1.67) + 168(s+1.67) + 140K_f s^2}\ . \qquad \ldots(6.59)$$

Multiplying out and collecting terms:

$$\frac{C}{R}(s) = \frac{168(s+1.67)}{s^4 + 7.67s^3 + 18s^2 + 181.4s + 280.6 + 140K_f s^2}\ . \qquad \ldots(6.60)$$

The characteristic equation can now be rearranged to the root locus form with $140K_f$ as the parameter.

$$1 + KG(s) = 1 + \frac{140K_f s^2}{s^4 + 7.67s^3 + 18s^2 + 181.4s + 280.6}\ . \qquad \ldots(6.61)$$

Before proceeding with the required locus of the system poles, the poles of $G(s)$ must be found by the partition method. Rewriting the quartic expression and equating to 0;

$$s^4 + 7.67s^3 + 18s^2 + 181.4\,(s + 1.55) = 0$$

or,

$$1 + \frac{181.4\,(s + 1.55)}{s^2\,(s^2 + 7.67s + 18)} = 0 \qquad \ldots(6.62)$$

The root locus obtained from (6.62) is shown in Fig. 6.16. The points on the four branches of the locus corresponding to a K of 181.4 are respectively, $s = -1.65$, $s = -7.6$, $s = +0.88 + j4.55$ and $s = +0.88 - j4.55$.

Equation (6.61) may now be rewritten:

$$1 + \frac{140K_f s^2}{(s + 1.65)\,(s + 7.6)\,(s - 0.88 - j4.55)\,(s - 0.88 - j4.55)} = 0 \qquad \ldots(6.63)$$

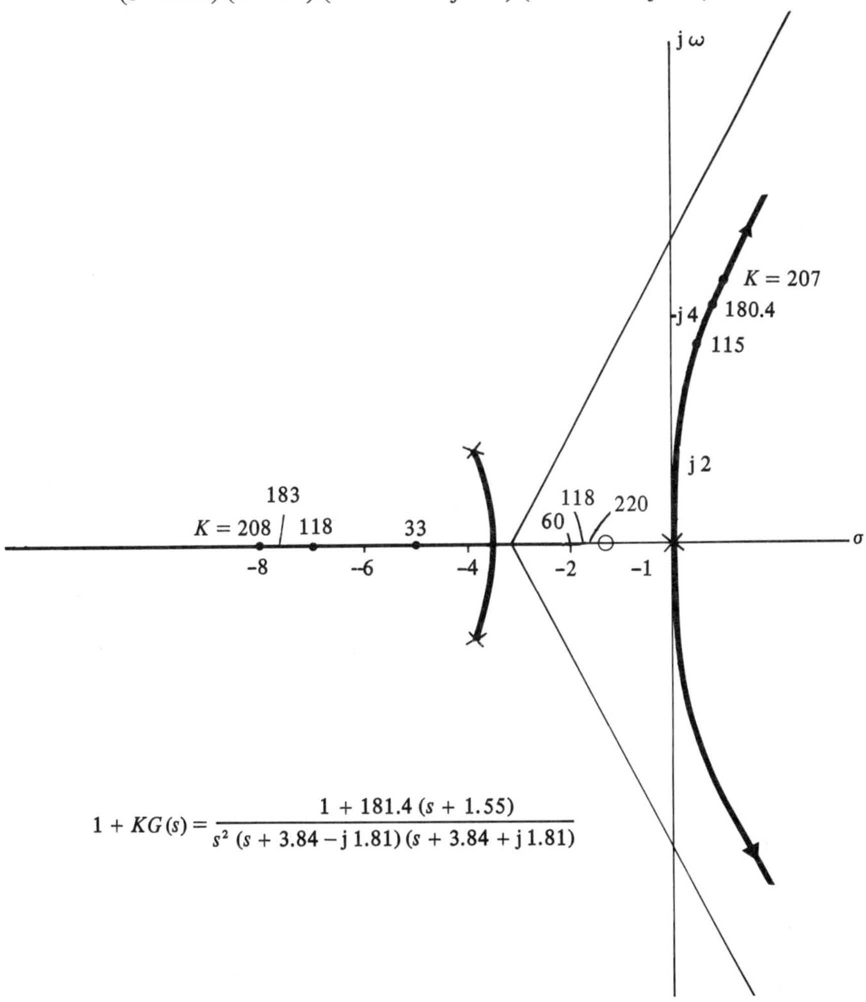

Fig. 6.16 Determination of the roots of a quartic expression by means of partition and root locus plot for Example 6.7.

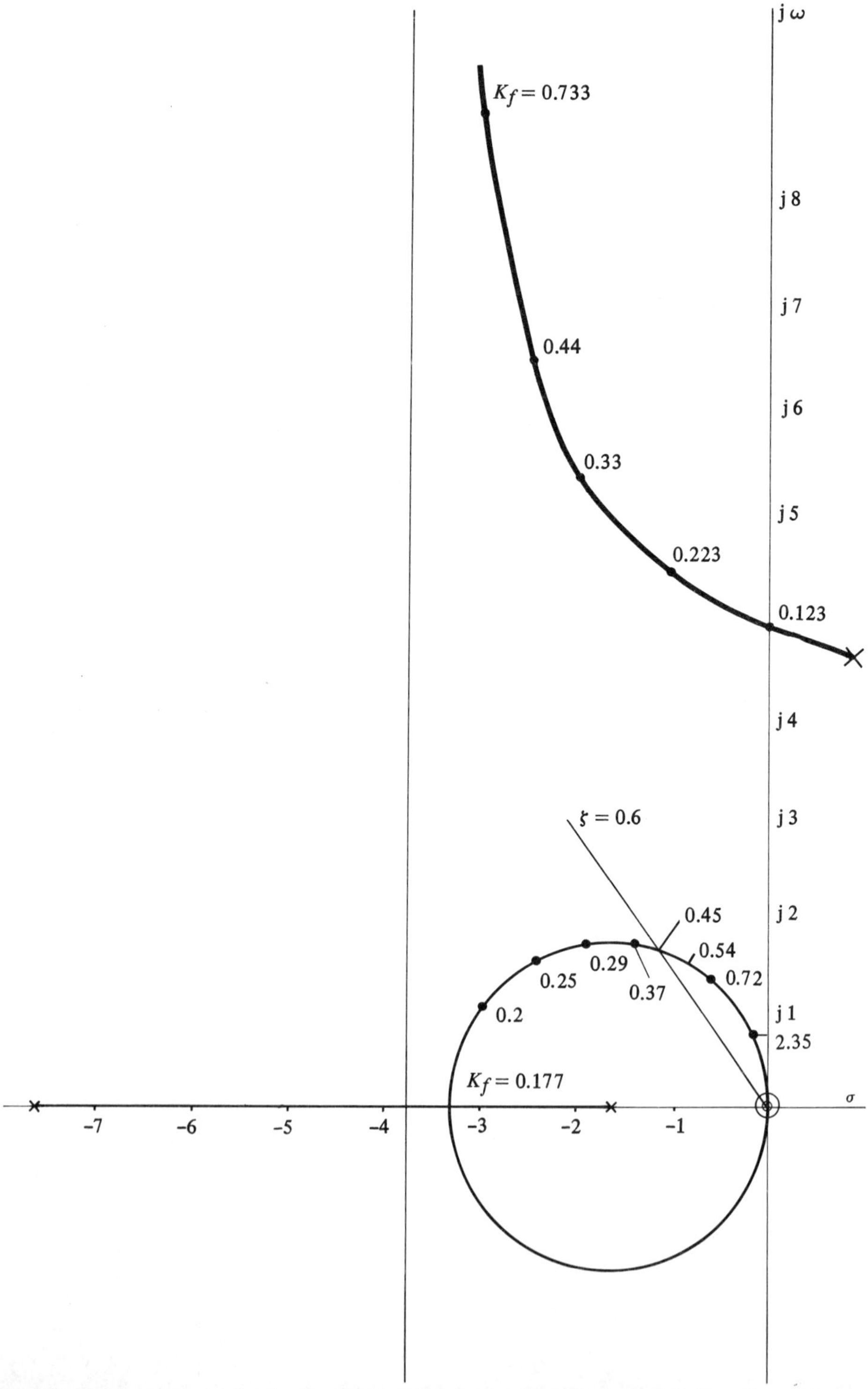

Fig. 6.17 The root locus plot for the automatic control system in Example 6.7. The effect of feedback compensation on the system poles.

From this result, the required locus is plotted as shown in Fig. 6.17. The scaling figures shown are for K_f, the parameter under investigation.

It is interesting to note that the unstable complex poles of $G(s)$ become more stable as K_f is increased, while the real poles of $G(s)$ become complex and approach the unstable condition at very low frequencies with high values of K_f.

Returning to the original question, a damping factor of 0.6 was specified for the dominant poles. The corresponding angle θ is $\cos^{-1} 0.6$ or 53°. This is shown in Fig. 6.17 and the resulting value of K_f is 0.45. The four poles can be read from the locus and by referring to expression (6.60), the system transfer function is obtained.

$$\frac{C}{R}(s) = \frac{168\,(s + 1.67)}{(s + 1.25 - j1.6)\,(s + 1.25 + j1.6)\,(s + 2.5 - j7.5)\,(s + 2.5 + j7.5)} \qquad \ldots (6.64)$$

The step response for this transfer function is shown in Fig. 6.18. It should be noted that the overshoot is more than might be expected for a damping factor of 0.6. This is the effect of the system zero and the second pair of complex poles. These high frequency poles also tend to 'speed up' the leading edge of the response, but there is a noticeable ripple.

Other possibilities which could be investigated are a change in K_a before plotting the locus for K_f or alternatively, a different value of K_f on the locus shown in Fig. 6.17. It would provide useful practice if the reader obtained the step response for a K_f of 0.37.

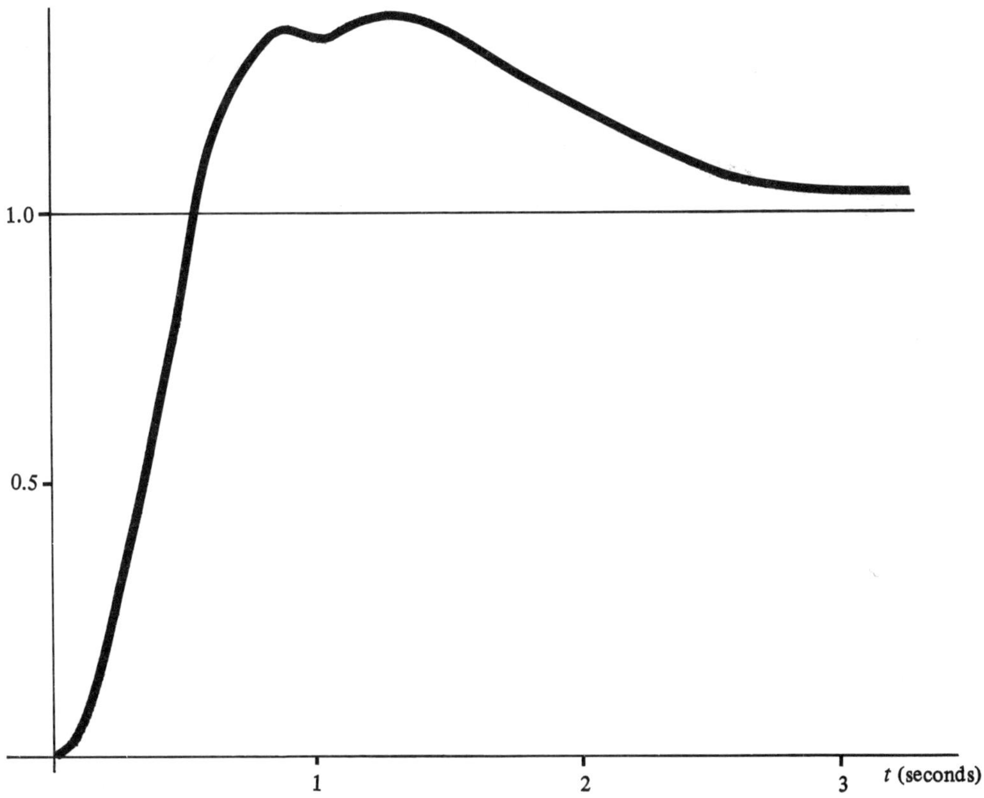

Fig. 6.18 The step response of the control system in Example 6.7.

6.8 ANALYSIS OF A SINUSOIDAL OSCILLATOR

In all our previous examples, the implication has been that instability in a system is undesirable. With sinusoidal oscillators, a steady-state sine wave output is required with no input. This would be obtained with a system having a pair of imaginary poles at $\pm j\omega$ where ω is the required frequency of oscillation. In practice, the poles will have positive real parts for small signals but as the amplitude of the output grows, the system gain will get smaller. An equilibrium condition will then exist with a steady oscillatory output. A root locus plot can be used to investigate choice of component values in an oscillator circuit as illustrated in the next example.

Example 6.8. A circuit for a Colpitt's oscillator is shown in Fig. 6.19. The transistor input resistance and current gain respectively (including the effect of the bias network) are 10 kΩ and 20. By means of a root locus, determine the range of C for which oscillation will take place and the resulting frequency range.

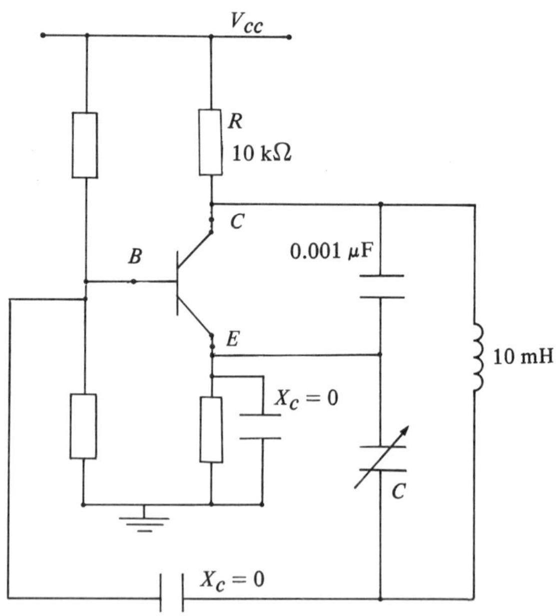

Fig. 6.19 The Colpitt's oscillator circuit for Example 6.8.

Solution. The small signal equivalent circuit is shown in Fig. 6.20a with the feedback path opened at the collector of the transistor. For analysis, I_2 can be found in terms of I_1; then, allowing for the current gain, the loop can be closed as a unity feedback system. Analysis can be simplified if the components are first impedance and frequency scaled by 10^3 and 10^5 respectively.

The resulting components are 10 Ω, 1 H, and 0.1 F. The redrawn circuit in Fig. 6.20b shows these scaled or normalized values expressed as admittances in the s domain.

The nodal equations are:

$$I_1 = V_1\,(0.1s + 0.1 + 1/s) - V_2/s \;; \qquad \ldots(6.65)$$

$$0 = -V_1/s + V_2\,(sC + 0.1 + 1/s)\;. \qquad \ldots(6.66)$$

From which:

$$V_2 = \frac{10I_1}{s^3C + (0.1 + C)s^2 + (1.1 + 10C)\,s + 2}\;. \qquad \ldots(6.67)$$

Since $I_2 = 0.1V_2$, the forward gain can be written:

$$A(s) = \frac{-20}{s^3C + (0.1 + C)\,s^2 + (1.1 + 10C)\,s + 2}\;. \qquad \ldots(6.68)$$

Substituting this result into the unity feedback form and rearranging:

$$A_f(s) = \frac{-20}{s^3C + (0.1 + C)s^2 + (1.1 + 10C)\,s + 22}\;. \qquad \ldots(6.69)$$

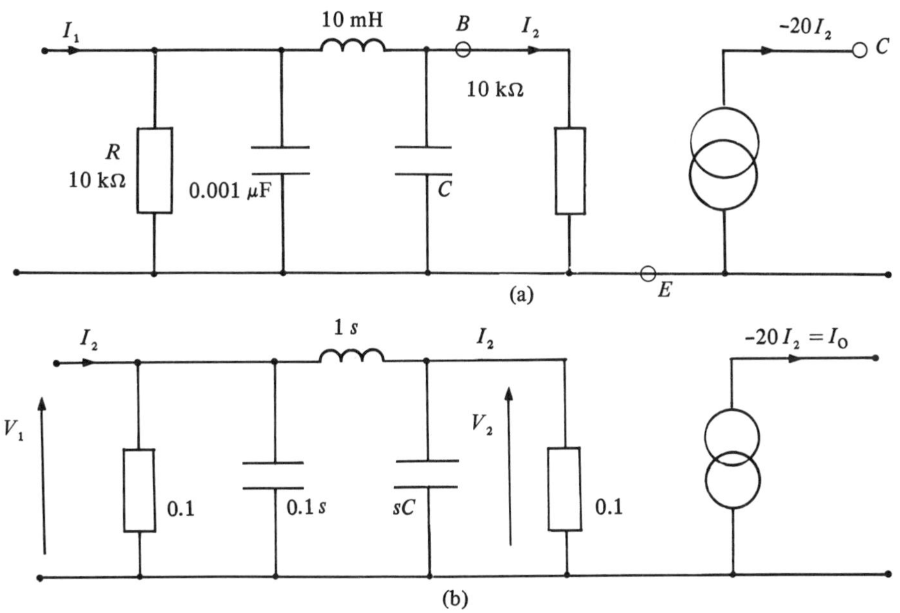

Fig. 6.20 Small signal equivalent circuits for the oscillator in Example 6.8. a) The equivalent circuit obtained from the full circuit with the feedback path broken. b) The frequency and impedance scaled version of (a) in the s domain.

From which, the *CE* is:

$$0.1s^2 + 1.1s + 22 + Cs\,(s^2 + 1s + 10) = 0\;, \qquad \ldots(6.70)$$

or, in the root locus form,

$$1 + \frac{10Cs\,(s^2 + s + 10)}{s^2 + 11s + 220} = 0\;. \qquad \ldots(6.71)$$

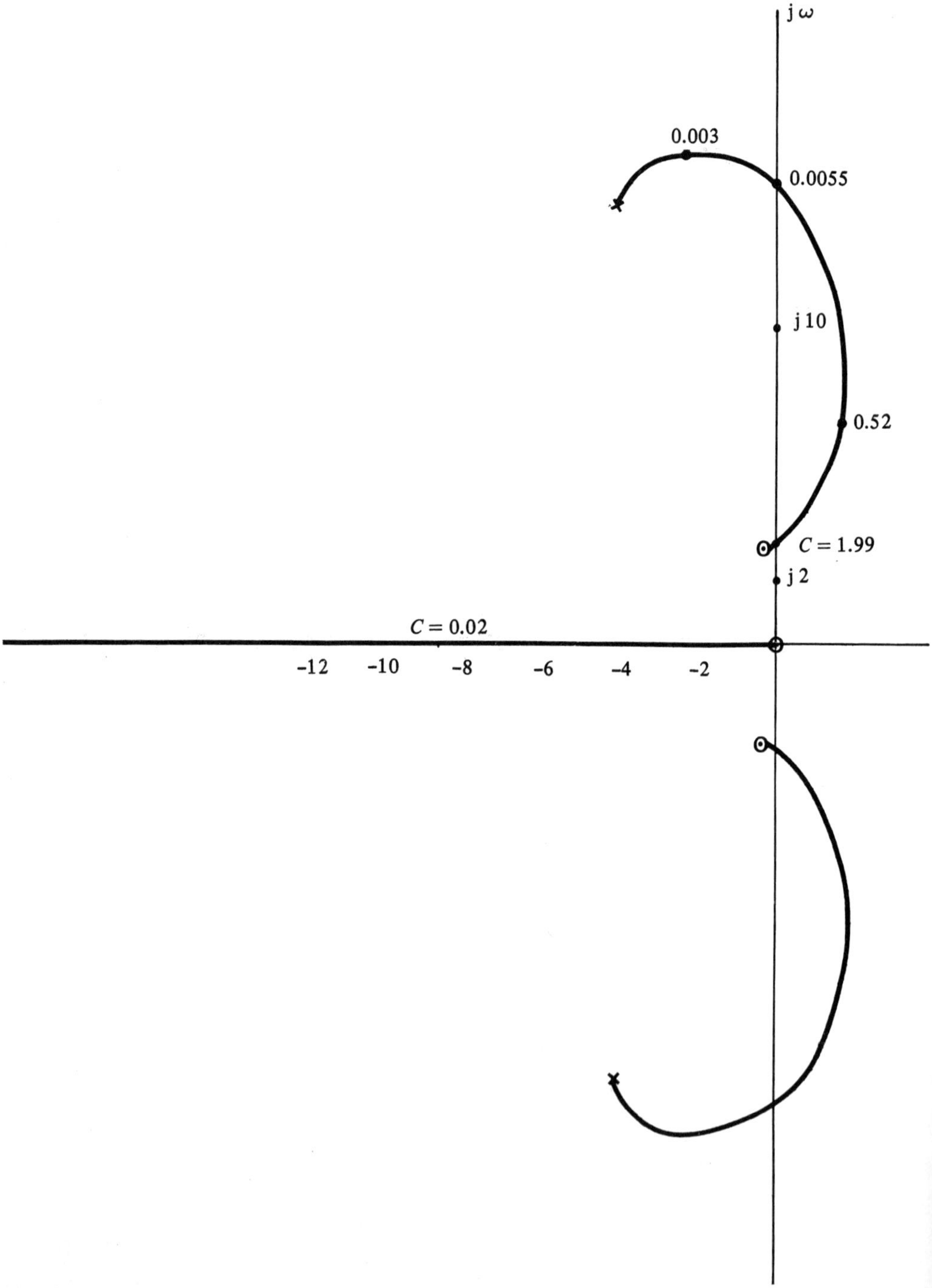

Fig. 6.21 A root locus plot showing the effect of the tuning capacitor on the pole positions for the Colpitt's oscillator in Example 6.8.

The resulting root locus is shown in Fig. 6.21. It is interesting to note this is an example where $G(s)$ has more zeros than poles. As a result, the single asymptote comes *from* a pole at infinity to the zero at the origin. The parameter for the locus is $10C$, but the scaling figures shown are for the C values.

The range of C for oscillation is 0.0055 F to 1.99 F resulting in a frequency range from 14.6 rads^{-1} down to 3.2 rads^{-1}. These values are of course for the normalized or scaled components. Converting back to the original circuit, the capacitor values are 55 pF and 0.0199 μF and the corresponding frequencies are 1.46×10^6 rads^{-1} and 0.32×10^6 rads^{-1}.

In this final chapter, examples have been used to illustrate some of the applications of root locus techniques. The root locus plots have been linked with system pole zero diagrams and the resulting transient and steady-state responses.

6.9 EXAMPLES FOR FURTHER PRACTICE

Example 6.9. Two unity feedback control systems have forward transfer functions given by: $T_1(s) = \dfrac{K}{s(s^2+6s+25)}$ and $T_2(s) = \dfrac{K(s+1)}{s(s+1.4)(s+2.4)(s+8)}$. In each case, apply root locus analysis to determine the system transfer function and hence the step response, if the complex poles have a damping factor of 0.4.

[$\dfrac{44}{(s+2.75)(s^2+3.24s+16.2)}$, $1 - 1.09e^{-2.75t} + 0.78e^{-1.62t}\sin(3.68t - 187°)$;

$\dfrac{50.5(s+1)}{(s+0.91)(s+8.9)(s^2+2s+6.29)}$, $1 - 0.124e^{-0.91t} - 0.083e^{-8.9t} + 1.063e^{-t}\sin(2.3t - 132°)$]

Example 6.10. A d.c. amplifier has a gain of 2000 and four high frequency breaks at 10^4, 2×10^5, 4×10^5 and 10^7 rads^{-1} respectively. If feedback, negative at low frequencies, is applied, determine the maximum feedback factor β for stable operation and the transfer function when β is one third of the maximum value. Hence construct the Bode gain plot for the feedback amplifier under these conditions.

[0.03; $\dfrac{1.6 \times 10^{25}}{(s+5.2\times10^5)(s+10^7)(s^2+10^5 s+3.313\times10^{10})}$; 39 dB with a resonant rise to 44.3 dB at 1.8×10^5.]

Example 6.11. An RC coupled amplifier has a mid-band gain of 1400, low frequency breaks at 100, 400 and 500 rads^{-1} and high frequency breaks at 10^5, 3×10^5 and 6×10^5 rads^{-1}. Feedback, negative at mid-band, is applied to improve the bandwidth. Taking a maximum Q factor of 1 for any complex poles, determine the feedback factor β and, by means of a Bode gain plot estimate the resulting amplifier bandwidth and mid-band gain. (1.2×10^{-3}, 280 to 2.5×10^5, 522.)

Example 6.12. The three stages of a d.c. amplifier each have a gain of 125 and a single pole at -10^6 rads^{-1}. Negative feedback is applied over the last two stages with a factor β_1 and over all three stages with a factor β_2. Taking the value of β_2 as 10^{-4}, determine the minimum value for β_1 and the frequency of oscillation if β_1 is just below this value. (6.3×10^{-3}, 10^7.)

Example 6.13. Determine the roots of the following polynomials using the partition and root locus method.

a) $s^3 + 15s^2 + 66s + 80$.

(–2, –5, –8.)

b) $s^3 + 11s^2 + 265s + 723$.

(–3, –4 + j15, –4 – j15.)

c) $s^4 + 13s^3 + 55s^2 + 111s + 140$.

(–4, –7, –1 + j2, –1 – j2.)

d) $s^5 + 15s^4 + 312s^3 + 1816s^2 + 3867s + 2169$.

(–1, –3, –3, –4 + j15, –4 – j15.)

Example 6.14. By means of a root locus plot, investigate the effect of the value of R_2 upon the transfer function poles for the compensator circuit shown in Fig. 6.22. Hence, find the value of R_2 if the high frequency pole is to be at –7000, or at –70 000 rads^{-1}. In each case, find also the remaining pole and zero positions and sketch the Bode gain and phase plots. (36.2 kΩ, –316, –4000, –552; 1.8 kΩ, –635, –4000, –11 100.)

Example 6.15. By means of a root locus plot, investigate the effect of the capacitor C upon the poles and zeros of the transfer function for the scaled network shown in Fig. 6.23. Hence, find the values of C to obtain a Q factor of a) 1 and b) 3. (1.44, 0.07.)

Example 6.16. Fig. 6.24 shows the circuit for a high Q band pass filter. Show that the transfer function is given by:

$$T(s) = \frac{sC_3G_1\ (1 + R_a/R_b)}{s^2C_3C_4 + s\,(G_5\ (C_3 + C_4) - R_a/R_bC_3G_1) + G_1G_5}\ .$$

Plot a root locus showing how R_a controls the Q factor for the values given. (125, $Q = 10$, 187.5, $Q = 20$, 250 unstable.)

Example 6.17. A servomechanism is employed to control the angular position of a mass having inertia 10^{-5} kg m^2 and viscous friction 1.5×10^{-4} Nm/rads^{-1}. The controller constant is set at 0.073 Nm/rad and the transient response is modified by positive acceleration and negative velocity feedback with constants of K_a and K_v respectively. Construct root locus plots showing how the system pole positions vary with K_a ($K_v = 0$) and with the value of K_v (K_a constant). (As an example, if $K_a = 1.11 \times 10^{-4}$, K_v must be increased to 3×10^{-3} to provide a damping factor of 0.5.)

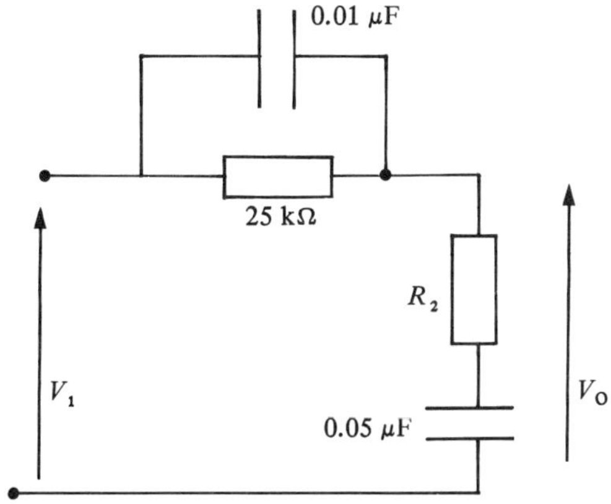

Fig. 6.22 Compensator circuit for Example 6.14.

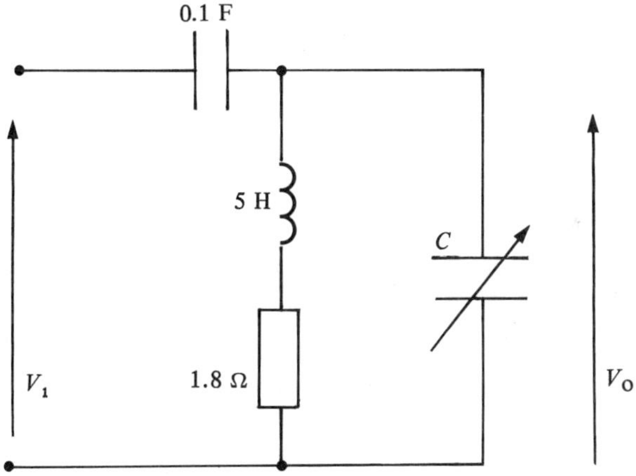

Fig. 6.23 Scaled network for Example 6.15.

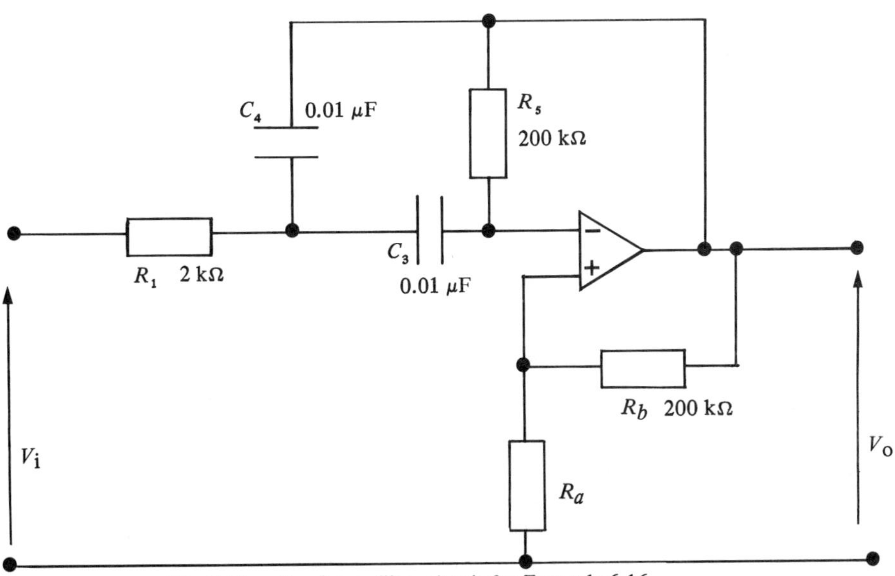

Fig. 6.24 Band pass filter circuit for Example 6.16.

Example 6.18. A type 2 unity feedback control system has a forward transfer function given by: $T(s) = \dfrac{K}{s^2\,(s+22)\,(s+33)}$. Compensation is to be provided by the addition of a single real zero. Investigate the optimum position for the zero assuming that speed of response with overshoot not exceeding 20 per cent is an important consideration. (If the zero is near the origin, the best result is obtained. For example, if z is at -2 and $K = 1.45 \times 10^4$,

$$T(s) = \frac{1.45 \times 10^4\,(s+2)}{(s+2.2)\,(s+47)\,(s+3.5-\mathrm{j}16)\,(s+3.5+\mathrm{j}16)}.$$

If the zero is at -5, two values of K produce 20 per cent overshoot but the lower value gives a much slower response. If the zero is at -6 or more, all values of K result in very light damping and a slow response.)

Example 6.19. A type 0 unity feedback control system has a forward transfer function given by: $T(s) = \dfrac{K}{(s+0.2)\,(s+0.5)}$. The performance is to be improved by the addition of a three-term controller which has a transfer function of $\dfrac{8.33\,(s^2+0.12s+17)}{s}$. Show that the resulting system is conditionally stable and determine the range of K for stability. Discuss the transient response obtaining to each stable range of K. (K less than 5×10^{-4}, sluggish response ringing at a frequency less than 0.05 Hz; K greater than 16.9, fast response but with a lightly damped oscillation at about 0.6 Hz.)

Example 6.20. Construct a root locus for the system having the block diagram shown in Fig. 6.25. Hence, find the maximum K value for stable operation. (279, at j7.45.)

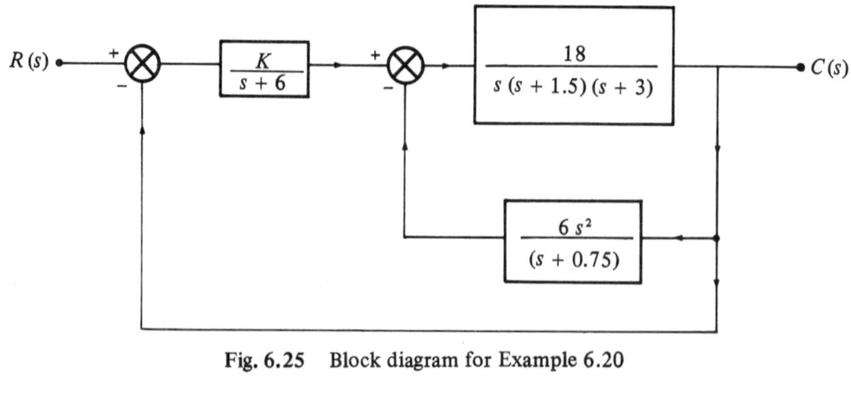

Fig. 6.25 Block diagram for Example 6.20

Example 6.21. The circuit shown in Fig. 6.26 is that for a simple phase shift oscillator. The transistor current gain and input resistance are 200 and 8 kΩ respectively. Neglecting the effect of the bias resistor R_B, determine the tuning range for the capacitor C and the resulting range of oscillatory frequency. (Root locus from $1 + \dfrac{0.67Ks\,(s^2 + 34.5s + 9852)}{s^2 + 1333s + 3.33 \times 10^5}$.

Tuning range 0.24 μF to 1.44 nF; 17 Hz to 85 Hz.)

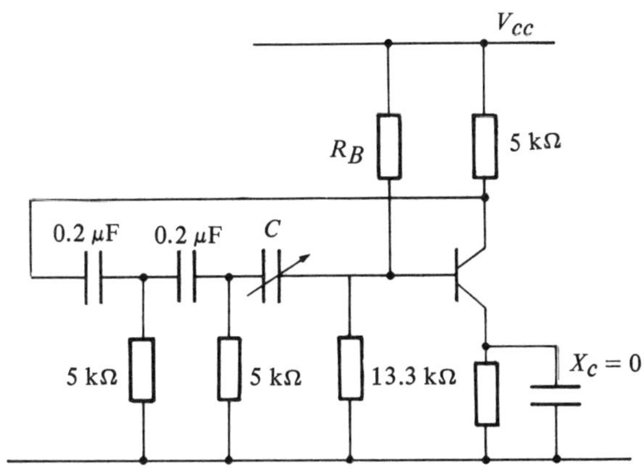

Fig. 6.26 Phase shift oscillator circuit for Example 6.21.

Appendix

A1 PARTIAL FRACTION EXPANSION

Analysis of electrical and control systems by use of Laplace transform methods requires the finding of the inverse transform from a result which, in general, is in the form of the ratio of two polynomials in s. The usual procedure is to divide the result into partial fractions which can be individually recognized as the transforms of known time functions. The purpose of this appendix is to summarize this technique of partial fraction expansion.

The first step is to factorize the denominator polynomial: each of the resulting factors will have one of the following forms.

i.	$(s + a)$	a simple non-repeated factor (including $s + 0$)
ii.	$(s + a)^n$	simple repeated factors (including s^n)
iii.	$s^2 + bs + c$	quadratic factors having complex roots (including $s^2 + c$)

For each simple unrepeated factor, there will be a single partial fraction term.

Example A1.1.

$$\frac{500}{s\,(s+1)\,(s+3)} = \frac{A}{s} + \frac{B}{s+1} + \frac{C}{s+3}.$$

For each simple factor raised to the power n, there will be n partial fraction terms.

Example A1.2.

$$\frac{3s+4}{(s+2)^3\,(s+1)} = \frac{A}{(s+2)^3} + \frac{B}{(s+2)^2} + \frac{C}{(s+2)} + \frac{D}{(s+1)}.$$

Note that the simple factor still has a separate term as before. For each quadratic term with complex roots, the partial fraction numerator will be of the form $As + B$.

Example A1.3.

$$\frac{500\,(s+2)}{s(s^2+2s+5)\,(s^2+9)} = \frac{A}{s} + \frac{Bs+C}{s^2+2s+5} + \frac{Ds+E}{s^2+9}.$$

Note that it may be advantageous to obtain the complex factors and to treat them as simple unrepeated factors. The last example can then be rewritten:

$$\frac{500\,(s+2)}{s\,(s+1-\mathrm{j}\,2)\,(s+1+\mathrm{j}\,2)\,(s-\mathrm{j}\,3)\,(s+\mathrm{j}\,3)} = \frac{A}{s} + \frac{B}{(s+1-\mathrm{j}\,2)} + \frac{B^*}{(s+1+\mathrm{j}\,2)}$$
$$+ \frac{C}{(s-\mathrm{j}\,3)} + \frac{C^*}{(s+\mathrm{j}\,3)}.$$

The remaining problem is the determination of the partial fraction *residues* A, B, C, etc. The residues for the simple unrepeated factors may, in all cases, be found by the cover-up rule.

The cover-up rule is applied in the following way:

i. Equate the expression to the partial fraction form described above.
ii. For each term in turn, determine the *value* of s which will make the denominator of that particular term equal zero.
iii. Substitute this *value* of s into the full expression (numerator and denominator) with the factor in question ignored or 'covered up'. Evaluating then results in the particular partial fraction numerator or *residue.*

Applying this procedure to Example A1.1 above, and rewriting:

$$\frac{500}{s\,(s+1)\,(s+3)} = \frac{A}{s} + \frac{B}{s+1} + \frac{C}{s+2}.$$

To find A; the denominator of the A term is simply s and this will be zero if $s = 0$. Substitution in the full expression with the s factor 'covered up' results in:

$$\frac{500}{(0+1)\,(0+3)} = 167 = A.$$

A This is often written $A = \left.\dfrac{500}{(s+1)\,(s+2)}\right|_{s=0} = 167.$

To find B; the term is $(s+1)$ which will be zero if $s = -1$. Hence,

$$B = \frac{500}{(-1)\,(-1+3)} = -250.$$

This is often written:

$$B = \left.\frac{500}{s\,(s+3)}\right|_{s=-1} = -250.$$

To find C, therefore:

$$C = \left.\frac{500}{s\,(s+1)}\right|_{s=-3} = \frac{500}{(-3)\,(-3+1)} = 83.3.$$

This method will also find the first residue for a repeated root. For the remaining residues, both sides of the equation are multiplied by the polynomial denominator and, having substituted for any known residues, one of the following techniques is applied.

i. Substitute a convenient numerical value for s (such as $s = 0$ or $s = -1$) so that certain terms will be eliminated.
ii. Equate coefficients of s^n to form a set of simultaneous equations which may then be solved to find the required residues.
iii. Multiply both sides by s and then let $s \to \infty$.

Applying some of these methods to Example A1.2 above:

$$D = \frac{-3+4}{(1)^3} = 1 \quad \text{and} \quad A = \frac{-6+4}{-1} = 2. \quad \text{(cover-up rule)}$$

$$3s + 4 = 2(s+1) + B(s+1)(s+2) + C(s+2)^2(s+1) + (s+2)^3$$

Let $s = 0$,

$$4 = 2 + 2B + 4C + 8$$

or

$$-3 = B + 2C$$

Equating coefficients of s^3,

$$0 = C + 1.$$

$$\therefore \quad C = -1 \quad \text{and} \quad B = -1.$$

A2 MESH AND NODAL EQUATIONS

The solution of complicated electrical networks is usually accomplished by either mesh or nodal analysis. Both of these methods are based on Kirchhoff's laws. For mesh analysis, a number of unknown currents are specified and then equations are written by equating volt drops around a loop to zero. For nodal analysis, a number of unknown node voltages are specified and equations are written by equating currents entering and leaving a node. In each case, if a convenient convention is used to specify the unknown variables, the necessary equations can be written by the application of a simple rule.

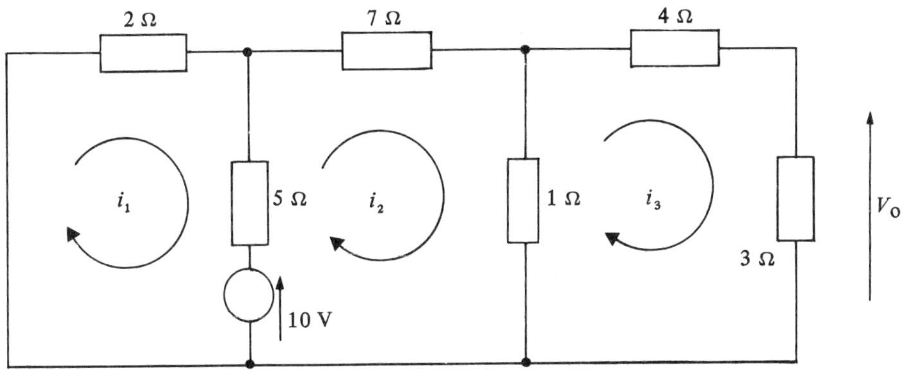

Fig. A.2.1 Impedance network for mesh analysis.

A2.1 MESH ANALYSIS

Specify the unknown currents by a circulating current in each loop or mesh of the network as shown in the impedance network in Fig. A2.1. Write the equations by the use of the following rule. For each loop, equate any e.m.f.s, taken in the direction of the loop current, to that loop current times (the sum of all the impedances around the loop), minus each adjacent loop current times (the common or shared impedance).

Applying this rule to the three loops in the circuit shown in Fig. A2.1,

loop 1,

$$-10 = i_1(2+5) - i_2(5);$$

loop 2,

$$+10 = -i_1(5) + i_2(5+7+1) - i_3(1);$$

loop 3,

$$0 = -i_2(1) + i_3(1+4+3).$$

The resulting simultaneous equations may then be solved by substitution and elimination or by the determinant method which is described below. A particular output voltage is then obtained from the appropriate unknown current impedance product.

A2.2 NODAL ANALYSIS

For nodal analysis, the network branches are more conveniently expressed as admittances and the specified *unknowns* are the voltages at the nodes or junctions between three or more branches. One node is chosen as the reference (usually the earth line), and after solution, the remaining node voltages will be expressed with reference to this chosen node.

Equations are formed by equating the currents entering a node from generators to currents leaving the node through admittances. There may be current generators in the network but more often (particularly in the determination of transfer functions) the applied signal is a *known* voltage in series with an admittance.

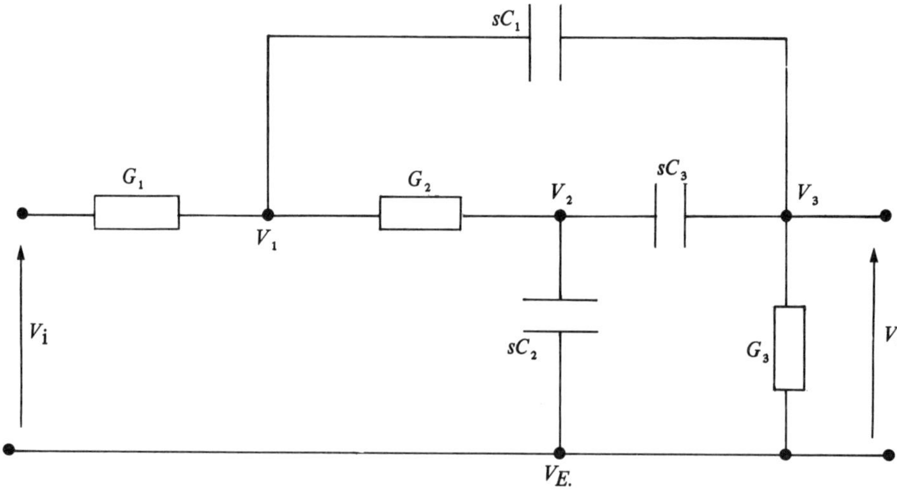

Fig. A.2.2 Admittance network for nodal analysis.

Consider the circuit shown in Fig. A2.2 for which the transfer function $\dfrac{V_o}{V_i}$ is required. V_E is the reference and V_1, V_2, V_3 are the specified unknown node voltages (note, $V_3 = V_o$).

V_i is a known or applied voltage. The equations can now be written for each unknown node by application of the following rule: For each node, equate any currents entering the node from a generator to that node voltage times (the sum of all the admittances connected between that node and the other unknown nodes) minus each adjacent unknown node times (the connecting admittance).

Applying this rule to the circuit in Fig. A2.2,

node 1,

$$(V_i - V_1)\,G_1 = V_1\,(G_2 + sC_1) - V_2G_2 - V_3sC_3\,,$$

node 2,

$$0 = -V_1G_2 + V_2\,(G_2 + sC_2 + sC_3) - V_3sC_1\,,$$

node 3,

$$0 = -V_1sC_1 - V_2sC_3 + V_3\,(G_3 + sC_1 + sC_3).$$

Note that the term $(V_i - V_1)G_1$ in the first equation is the current from the known voltage generator.

Once again, we have a set of simultaneous equations to be solved, but in this case, a solution for V_3 will lead directly to the required transfer function.

A2.3 SOLUTION BY DETERMINANTS

A determinant is an array of terms having an equal number of rows and columns, this number is known as the order of the determinant. For example,

$$D_1 = \begin{vmatrix} 3 & 5 \\ 2 & 8 \end{vmatrix} \text{ is a second-order determinant;}$$

and

$$D_2 = \begin{vmatrix} 1 & 0 & -1 \\ 2 & s & s \\ 2s & 1 & 0 \end{vmatrix} \text{ is a third-order determinant.}$$

If determinants are to be used for the solution of a set of simultaneous equations, determinants made up from the coefficients of the unknowns together with any constants must be *expanded.*

The second-order D_1 is expanded as follows:

$$D_1 = (3 \times 8) - (5 \times 2) = 14.$$

For a third order, the procedure is a little more complicated but may be illustrated for D_2 as follows:

$$D_2 = \begin{vmatrix} \textcircled{1} & 0 & -1 \\ 2 & \boxed{s} & \boxed{s} \\ 2s & \boxed{1} & \boxed{0} \end{vmatrix} - \begin{vmatrix} 1 & \textcircled{0} & -1 \\ \boxed{2} & s & \boxed{s} \\ \boxed{2s} & 1 & \boxed{0} \end{vmatrix} + \begin{vmatrix} 1 & 0 & \textcircled{-1} \\ \boxed{2} & \boxed{s} & s \\ \boxed{2s} & \boxed{1} & 0 \end{vmatrix}$$

The circled term in each case is multiplied by the expanded determinant in the box. Note also that the sign of the circled term alternates (+, –, +).

$$D_2 = 1\,(s \times 0 - s \times 1) - 0\,(2 \times 0 - 2\,s \times s) + (-1)\,(2 \times 1 - 2\,s \times s) = 2s^2 - s - 2.$$

If the technique is extended to fourth- or higher-order determinants, the boxes themselves will be third-order and so on.

Returning now to Example A2.1, the equations may be rewritten:

$$-10 = 7i_1 - 5i_2 + 0i_3,$$

$$+10 = -5i_1 + 13i_2 - 1i_3,$$

$$0 = 0i_1 - 1i_2 + 8i_3.$$

The network determinant is that made up from the coefficients of the unknowns.

$$D = \begin{vmatrix} 7 & -5 & 0 \\ -5 & 13 & -1 \\ 0 & -1 & 8 \end{vmatrix} = 7(104 - 1) + 5\,(-40) + 0 = 521.$$

When a particular unknown is required (in this case i_3), a second determinant is written with the coefficients of the required unknown replaced by the column of constants. For i_3, the required determinant D' is given by:

$$D' = \begin{vmatrix} 7 & -5 & -10 \\ -5 & 13 & +10 \\ 0 & -1 & 0 \end{vmatrix} = 7(+10) + 5(0) + (-10)(5) = 20.$$

The unknown current i_3 is then given by the ratio

$$\frac{D'}{D} = \frac{20}{521}.$$

Finally,

$$V_o = 3i_3 = \frac{60}{521}.$$

Applying the same technique to the second example;

$$V_iG_1 = V_1(G_1 + G_2 + sC_1) - V_2G_2 - V_3sC_1,$$

$$0 = -V_1G_1 + V_2(G_2 + sC_2 + sC_3) - V_3sC_3,$$

$$0 = -V_1sC_1 - V_2sC_3 + V_3(G_3 + sC_1 + sC_3).$$

Expanding the determinants directly;

$$V_3 = \frac{0 + 0 + V_iG_1\,[G_1sC_3 + sC_1\,(G_2 + sC_2 + sC_3)]}{(G_1 + G_2 + sC_1)[(G_2 + sC_2 + sC_3)(G_3 + sC_1 + sC_3) - s^2C_3^2] + G_2[G_1(G_3 + sC_1 + sC_3) - s^2C_1C_3] - sC_1\,[-sC_3G_1 + sC_1(G_2 + sC_2 + sC_3)]}.$$

This result could, of course, be simplified or evaluated for numerical values of the components.

A3 THE SCALING OF ELECTRICAL CIRCUITS

When a general analysis of a particular circuit form or a synthesis of a particular function is being attempted, it is usually convenient to work with very simple numerical values. This is often achieved by choosing a resonant or cut-off frequency of 1 rads^{-1} and a standard impedance level of 1 Ω. The resulting networks and components are convenient for comparison purposes but the components and impedance levels are not usually practical as they will be farads, henrys and ohms. This *normalized* circuit must be denormalized or scaled to obtain components that provide the required frequency performance and impedance levels. This is achieved in two stages: frequency scaling and impedance scaling.

A3.1 FREQUENCY SCALING

If L_0 and C_0 are the values of inductance and capacitance scaled to a frequency ω_0 and new values of L and C are required to be scaled to a frequency ω where, $\omega = \omega_0 \times n$,

Then,

$$L = L_0 \times \frac{\omega_0}{\omega} = \frac{L_0}{n},$$

and

$$C = C_0 \times \frac{\omega_0}{\omega} = \frac{C_0}{n}.$$

For example, a simple series circuit consists of a 2 H inductor, a 1 Ω resistor and a 0.3 F capacitor. This resonates at $\omega_0 = \frac{1}{\sqrt{0.6}}$ or 1.29 rads^{-1}. Also, X_L = 2.58 Ω, X_C = 2.58 Ω and Q = 2.58 at this frequency.

If the components are now frequency scaled to ω_1 = 1000 rads^{-1} i.e. scaled by a factor of 775, the scaled components are,

$$L = \frac{2}{775} = 2.58 \text{ mH and } C = \frac{0.3}{775} = 387\ \mu\text{F}.$$

These scaled components now resonate at

$$\frac{1}{\sqrt{2.58 \times 10^{-3} \times 387 \times 10^{-6}}} = 1000 \text{ rads}^{-1}.$$

The impedance levels are, however, unchanged:

$$X_L = 1000 \times 2.58 \times 10^{-3} = 2.58\ \Omega \text{ and } X_C = \frac{1}{1000 \times 387 \times 10^{-6}} = 2.58\ \Omega$$

and the Q factor is unchanged at 2.58.

Thus the frequency of operation of a circuit can be increased by a factor n if the values of all the inductors and all the capacitors are divided by n (a similar reduction would result from multiplication by the chosen factor). Resistors are of course unchanged by frequency scaling.

A3.2 IMPEDANCE SCALING

If the impedance of a circuit is to be changed by a factor m, inductors and resistors must be multiplied by m while capacitors are divided by m ($X_C = \frac{1}{\omega C}$). Thus, if the previous example is to be impedance scaled by a factor of 40, the three new components become:

$$L = 2.58 \times 40 = 103 \text{ mH},$$

$$R = 1 \times 40\,\Omega,$$

$$C = \frac{387}{40} = 9.68\,\mu\text{F}.$$

The resonant frequency is now given by

$$\frac{1}{\sqrt{0.103 \times 9.68 \times 10^{-6}}} = 1000 \text{ rads}^{-1}.$$

$X_L = X_C = 1000 \times 0.103 = 103\,\Omega$ at this frequency and $Q = \dfrac{103}{40} = 2.58$.

Thus a series resonant circuit having ω_0 1.29 radians per second, impedance at resonance of 1 Ω and Q factor 2.58 has been scaled to a resonant frequency of 1000 rads^{-1}, impedance at resonance of 40 Ω but with an unchanged Q factor of 2.58. The fact that the Q factor is unchanged shows that the circuit selectivity expressed as the ratio of resonant frequency to bandwidth is unchanged. This means that the frequency properties of the circuit are the same as those for the original normalized circuit.

Index